P9-ECK-163

INTERACTIONS

INTERACTIONS

Some Contacts between the Natural Sciences and the Social Sciences

I. Bernard Cohen

The MIT Press
Cambridge, Massachusetts
London, England

This book was printed and bound in the United States of America.

Library of Congress Cataloging-in-Publication Data

Cohen, I. Bernard, 1914–

Interactions : some contacts between the natural sciences and the social sciences / I. Bernard Cohen.

p. cm.

Includes bibliographical references and index.

ISBN 0-262-03223-6. — ISBN 0-262-53124-0 (pbk.)

1. Science—Social aspects—History. 2. Science—History. 3. Sociology—Methodology—History. I. Title.

Q175.5.C62 1994

303.48'3—dc20 94-11074

CIP

For Robert K. Merton and Thomas S. Kuhn
and (on a different plane of intimacy)
for Susan T. Johnson

CONTENTS

PREFACE

This volume explores certain historical interactions between the social sciences and the natural sciences.[1] My hope is that it will add some necessary perspective to the ever-growing body of literature on the logical, philosophical, and "scientific" foundations of the social sciences. Most discussions of these topics, both of social science in general and of the individual social sciences, have not been conceived in a historical mode. The result is that, despite some notable exceptions, most writers have tended to examine the methods of the social sciences by comparison and contrast with the methods of the physical and biological sciences *as they exist today*, while ignoring the historical encounters and interactions between social scientists and the natural sciences of their own day. (Throughout the present work, I use "natural sciences" to designate the physical and biomedical sciences, plus mathematics and the earth sciences.)

There is a rapidly growing literature concerning the history of the individual social sciences, and a major journal in this area, *Journal of the History of the Behavioral Sciences*, ably edited by Barbara Ross, is currently in its thirtieth volume. Yet most of the research and writing on the history of the social sciences, however valuable in its own terms, has tended to be either an examination of the internal development of the discipline or a study of the relations of one of the social sciences to the larger intellectual and social matrix. There have been very few studies having the goal of analyzing the development of the social sciences in relation to concurrent developments in the physical and biological sciences. Pitirim Sorokin's *Contemporary Sociological Theories* and Joseph Schumpeter's *History of Economic Analysis*, for example, are two very useful compendia that barely mention the physical or biological sciences. This lack is glaring in Sorokin's analysis of the nineteenth-century organismic sociologists, who in fact drew heavily on such current or then-recent developments in biology as the cell theory, the discoveries concerning embryological development in mammals, the physiology of the "milieu intérieur," and the germ theory of disease. The same deficiency is also conspicuous in Schumpeter's presentation of the founders of marginalist economics, who based their concepts and methods on those of rational mechanics. An extreme example of this lacuna is Werner Stark's important historical analysis, *The Fundamental Forms of Social Thought*, which contains many lengthy quotations dealing with advances in the biological sciences (e.g., the work of Rudolf Virchow) but has no discussion of these biological principles, no hint of their importance in the development of the natural sciences, and no suggestion of their significance as examples of interactions between the natural sciences and the social sciences; similarly, the lengthy extracts and descriptions of the use of physical science by social scientists are presented without any inquiry into their use as other than rhetoric. Even so insightful a contribution as Dorothy Ross's recent *The Origins of American Social Science* takes no real cognizance of the actual physical and biological sciences that were utilized by the social scientists whose careers she explores. To be sure, the purposes of these authors were very different from examining the interactions between the social sciences and the physical and biological sciences, but their books do indicate, in a dramatic way, the need to improve our understanding of how the social sciences and the physical and biological sciences have interacted in the centuries since the advent of "science" as we know it today.

During the last several years, some scholars concerned with the history of the social sciences have begun to take cognizance of their interactions with the natural sciences, and these writings have been valuable for my own investigations. In particular, I have drawn heavily on the writings of the historians of science Theodore Porter, Robert Richards, Judith Schlanger, George Stocking, and Norton Wise.[2] Work by a group of economists who are studying the foundations of neoclassical or marginalist economics in the physical sciences and biological sciences has also proved important in the context of this volume. This group includes Philip Mirowski, Roy Weintraub, Neil de Marchi, Claude Ménard, Bernard Foley, Margaret Schabas, Arjo Klamer,[3] and especially Giorgio Israel and Bruna Ingrao,[4] whose seminal work runs parallel to my own and has been very important in the development of my thinking about the natural sciences and the social sciences. Additionally, studies on statistics by Ian Hacking, Stephen Stigler, Lorraine Daston, William Coleman, Gerd Gigerenzer, Lorenz Krüger, and the Bielefeld group[5] have added new perspectives on the relationship of core techniques to social problems and social theory. Although the present work does not deal with anthropology, I want to note the important new historical work in this area, primarily the serial publication, founded and edited by George Stocking, called *History of Anthropology*.[6] I should like particularly to acknowledge the importance for my own thinking in this area of the writings of Theodore Porter, Norton Wise, Giorgio Israel, George Stocking, and Claude Ménard.

I have not attempted to cover all branches of the social sciences in this book. Some important kinds of interactions are barely mentioned or are not discussed at all. For example, psychology and anthropology are not discussed, nor is history. Political science appears primarily in the setting of the Scientific Revolution of the seventeenth century. An additional limitation is the exclusion of methodological writings, such as those of John Stuart Mill.

My interest in the interactions of the social sciences and the natural sciences grew out of previous research on scientific creativity, which had focused on the different ways in which the sciences have influenced one another. It was but a short step to extend this inquiry into the interaction of the natural sciences with the social sciences. When I first undertook this investigation, I naively believed that the vast literature on historical aspects of the social sciences would provide a useful and readily available, if not fully digested, body of reliable material to serve my purpose. The very existence of two multi-volume encyclopedias of the social sciences, replete with biographies and bibliographies and historical expositions of main themes, seemed a guarantee that I would not have to do all the research in primary sources that is almost always required in my own field of history of science. After all, I reasoned, the social sciences represent a proud ancient profession with a direct lineage that can be traced to Plato and Aristotle. I naively supposed that social scientists would have paid attention to the interactions of their disciplines with the natural sciences during the centuries since the Scientific Revolution![7]

I was aware, furthermore, that some social sciences (notably psychology, political science, economics, and sociology) regularly include courses in the history of their discipline in the undergraduate and graduate curriculum, and some also make creative use of texts of past great masters in their teaching and research. Surely, I felt, these educational tools would help ease my way.

Another factor that led me to suppose that my task would be easier than it turned out to be was the constant litany in the different social sciences—primarily economics and sociology—of their status as sciences. This quite naturally led me into the error of believing that, in their studies of the past, social scientists would have particularly stressed the different ways in which their predecessors had drawn on the work of contemporary natural scientists as well as philosophers and their fellow social scientists.

No sooner had I started my research than I discovered that I was mistaken in all my assumptions. There was little literature that took account of the ways in which social scientists of the past three centuries[8] had attempted to use the concepts, principles, theories, or methods of the natural sciences at large.

Additionally, the influence of the social sciences on the development of the natural sciences was all but completely ignored and in some cases even denied. The only major area in which such studies had been undertaken in any serious way was economics.[9]

I did not understand how this situation could exist until I happened to re-read Robert Merton's Introduction to the collection of his essays on *Social Theory and Social Structure*. In the course of this general prolegomenon, Merton makes an important distinction between "the *history* of sociological theory" and "the *systematics* of certain theories with which sociologists now provisionally work." The confusion of genuine historical investigation and the retrospective search for "utilizable sociological theory" influences much of the writing on the history of sociology. A paradigmatic example is a work to which I have already referred, Pitirim Sorokin's retrospective survey, *Contemporary Sociological Theories*. The stated purpose of the book is to provide background information on the current state of knowledge through analytical and critical summaries of the ideas of nineteenth-century and early twentieth-century pioneers. Sorokin's aim is not so much to understand the thought of the past as to criticize the writings of all previous ages from a "presentist" standpoint and to seek for any useful principles that might still be valid. Hence this work is more a contribution to practical sociological methodology than to historical inquiry.

Merton's analysis applies equally well to other social sciences. Much of the history of economics is conceived in relation to economic theory, as a subject of direct use in understanding or in teaching economics. The field tends to be dominated by a critical attitude that has come to be known as Whiggism in history: the attempt to judge the ideas of the past by present standards rather than to explore such ideas in their historical context. This aspect may be discerned in the fact that many of the works in this area are devoted to specialized topics that are of interest today rather than to the nature of the subject as it existed in some past age. There are, of course, important exceptions. Schumpeter's *History of Economic Analysis*, for example, is a highly personal statement drawing on first-hand knowledge and historical insight. One of the most interesting general histories of any of the social sciences, this great work sparkles with individual judgments based on the author's values and the state of economics at the time.

From a long historical point of view, the influence of the natural sciences on the social sciences is not a new phenomenon born of the Scientific Revolution but rather as old as the idea of science itself. In his *Politics,*[10] Aristotle recommended that the study of constitutions and "the forms of government" be modeled on the methods of classifying "the different species of animals."

According to Sir David Ross, Aristotle even attempted to "achieve for States" the same "precise description of their types as he gives for animals in the *Historia Animalium*."[11]

In the Middle Ages and the Renaissance, the idea of the body politic explained the functions of government by analogy with human anatomy and Galenic physiology. One survival, of many, from this physiological political theory is the concept of a "head" of state. In the seventeenth century, the discoveries of Harvey and the influence of Descartes transformed the notion of the political body into the more modern form with which we are familiar today.

In the Renaissance, Queen Elizabeth's power was displayed in a diagram modeled on the current astronomical diagrams of the system of celestial spheres. Elizabeth I (reigning in the "sphaera civitatis") became the prime mover of the system, with the inner spheres representing her virtues or "planetary" attributes: abundance, eloquence, clemency, religion, fortitude, prudence, and majesty.[12] The Scientific Revolution produced a modified astro-political diagram in which the royal power of Louis XIV was presented in a background of a Copernican or heliocentric rather than an Aristotelian or geocentric system of the universe. The cosmology had been upgraded by setting the planetary system in a nest of Cartesian vortices. The date of birth of the monarch was given as a basis for computing the royal horoscope.[13] Louis's designation as "roi soleil" invoked an analogy between celestial phenomena and political power, just as Harvey's analogy between the role of King Charles and the function of the heart drew on biological science. There was, clearly, a long-standing tradition of associating theories of the state or social organization with current conceptions of science.

I show in Chapter 2 that Hugo Grotius was a great admirer of Galileo and that he conceived his celebrated treatise on international law to have been written in the spirit and manner of a work on geometry. This aspect of his work is not mentioned in the article on Grotius in either the older *Encyclopaedia of the Social Sciences* (1932) or the more recent *International Encyclopedia of the Social Sciences* (1968). Yet his ideal of geometry is relevant to an evaluation of his work because this feature determined that he would deal with abstract cases rather than historical or contemporary examples—an aspect of his presentation for which he has been roundly criticized.

The situation is somewhat the same for James Harrington, whose politico-social thought, expressed in his *Oceana* and other writings, attracted significant attention in the eighteenth century, influencing many of the Founding Fathers of the United States and becoming embodied in the U.S. Constitution. Although Harrington expressly founded or justified his system on the basis of the new

Harveyan physiology, there is no mention of Harvey or his science in the *Encyclopaedia of the Social Sciences*. In the *International Encyclopedia of the Social Sciences*, Harvey's influence is mentioned in passing, but not in a way that would give the reader any sense of the extent of his actual influence on Harrington.[14]

An equally striking example is provided by Leibniz's early essay offering a mathematical demonstration of a method of selecting a king for Poland. I find it significant that this essay does not merit notice in standard presentations of the history of political thought. The essay is not even mentioned in a recent volume devoted to Leibniz's political writings.

Even when the scientific component of social thought is introduced, its significance may be lost because of a lack of understanding. An example discussed in Chapter 1 involves Berkeley's conception of a social analogue of the Newtonian gravitational cosmology. Berkeley's presentation shows that he understood perfectly the principles of Newtonian celestial dynamics, explaining planetary orbital motion as a combination of a continual central accelerating force and an undiminished initial component of linear inertial motion along a tangent. In Sorokin's survey of Berkeley's Newtonian sociology, Berkeley's correct physics is reduced to the incorrect form of a "balance" between centripetal and centrifugal forces, a standard textbook error that has plagued the teaching of physics. Berkeley's sound Newtonian physics is represented by Sorokin's statement that stability occurs when the alleged centrifugal force is less than the centripetal force. Berkeley would have known, as Sorokin evidently did not, that in such a hypothesized example the unbalanced centripetal force would produce instability, with a resulting accelerated motion of falling inward toward the sun or other center of force. A somewhat similar example is Henry Carey's model of a social analogue of Newton's gravitational physics, mentioned or discussed in almost every historical work on social theories that I have encountered. In not one is it recognized that Carey made a fundamental mistake in stating Newton's law, the basis of his social science.

A considerable literature exists on the organismic sociologists of the late nineteenth and early twentieth centuries, a company that includes J. C. Bluntschli, Paul von Lilienfeld, Albert Schäffle, Herbert Spencer, Lester Ward, Corrado Gini, Walter Bradford Cannon, A. Lawrence Lowell, and Theodore Roosevelt. All except Spencer are discussed in historical surveys of works on sociological theory without any reference to their use of the leading biological and medical theories of their time. This absence is the more remarkable in the degree to which some of these scholars (notably Lilienfeld, Schäffle, and Cannon) included extensive bio-medical tutorials in their sociological presentations. Thus, however extravagant the ideas of these organismic sociologists may seem to us today, our judgment should not be limited by present-day concepts and standards. Rather, we should judge these theories by the standards of their own time, taking note that their authors displayed a deep and thorough knowledge of the latest biological concept and theories.

One aspect of the interactions between the natural and the social sciences that is all but wholly absent from the literature of both the history of the social sciences and the history of the natural sciences is the impact of the social sciences on the biological and physical sciences. Three examples will indicate this "reverse" kind of interaction. It has long been known that Darwin was influenced by Malthus's ideas on population growth while formulating his concept of natural selection. We now also know, thanks to the research of S. S. Schweber, that Darwin's thinking was significantly influenced by the agronomists. Another idea that Darwin obtained from the social sciences, one that became particularly important during the nineteenth century in the context of the cell theory, was the division of labor. This concept gained prominence through the writings of Adam Smith, although the idea had been put forth earlier by such writers as William Petty and Benjamin Franklin. As we have learned from the research of Camille Limoges, this social concept of division of labor became particularly significant in the thought of the French biologist Edouard Brown-Séquard, who applied it in relation to the role of individual cells in the physiology of the organism. From him the concept was transmitted to Emile Durkheim, who wrote a major work on the sociological division of labor. Even more striking may be the case of statistics, where—as we know thanks to the research of Theodore Porter—there was an important influence of the Belgian social statistician Adolphe Quetelet on the physics of both Maxwell and Boltzmann.[15]

The general importance of Quetelet and the rise of statistical thinking in the social sciences may be seen as a special case of the interaction of mathematical techniques and social thought. The latter subject has attracted the attention of a number of scholars. We should keep in mind that the introduction of a statistical point of view aroused considerable alarm. Many thinkers—including John Stuart Mill and Auguste Comte—considered statistics the resort of an incomplete and faulty science that had failed to produce a simple Newtonian one-to-one relation between cause and effect. Comte not only pilloried Quetelet and others for adopting a statistical point of view but even gave up his original designation of "social physics" for what he then named "sociology" because "social physics" had been used in a probabilistic framework by Quetelet. Much of the subsequent development of social thought may be seen as reflecting a tension between the ideas of Comte and Quetelet, between a social science exhibiting simple cause and effect and one based on statistical considerations.

A major theme of the present volume is the role of analogies in the development of the social sciences. I shall propose a distinction between analogies and homologies and between both of these and metaphor, and I shall also focus on the problems that arise in using concepts, laws, or theories from the natural sciences in the social sciences. In the nineteenth century there were two notable examples. One was the development of a mathematical marginalist economics by Jevons, Walras, and Pareto on the model of rational mechanics plus energy physics; the other involved the use of the cell theory and certain allied aspects of biology and medicine by the organismic physiologists.

Because my goal has been to explore methodological issues, I have concentrated on selected main figures in the nineteenth-century debates concerning the modeling of society, that is, the question of whether society should be conceived as a physical mechanism or an organic being. I am fully aware that an encyclopedic survey would take account of such important pioneers as Machiavelli, John Locke, and Max Weber. I have similarly omitted Karl Marx, who claimed to have created a new "science" of society. Similarly, although I have introduced the ideas of Pareto in relation to his belief in the analogy between economics and rational mechanics, I have not had space to deal with his compatriot, the Italian economist, statistician, and sociologist Corrado Gini, who should be reckoned among the primary organicists.

Although known today primarily as an economist and statistician, Gini is also held to be one of the major twentieth-century advocates of an organicist theory of society.[16] In his organismic sociology, Gini not only developed general analogies (such as medical and social or economic pathology), but even "formulated a theory of social metabolism" that exhibits all the essential features of organic metabolism[17]—a striking instance of the use of the natural science by a social scientist.

Finally, anyone who studies the relationships between the natural and the social sciences becomes aware that this is not a purely academic topic but rather one that has close links to policy questions. First of all, the social sciences seek legitimation by the degree to which they resemble the natural sciences and actually incorporate features, concepts, laws, or theories of the natural sciences. Because most people think of physics when they consider what a science should be, social sciences are most impressive to the general public when they are based on extensive numerical foundations or exhibit mathematical procedures in the manner of physics. Public support of the social sciences under the umbrella of "science," as in the case of the National Science Foundation, will seem most appropriate—and may more readily become fact—for those social sciences that interact with or emulate the natural sciences. Such considerations are directly related to the natural scientists' images of the social sciences and were influential during the Congressional hearings on the establishment of the National Science Foundation.

In recent decades, natural scientists have expressed concern about the present state and future needs of the social sciences. This broad subject is relevant to the main assignment of the present volume, although it is far too complex to be incorporated into a single chapter. Accordingly, a series of interviews conducted by the author with Harvey Brooks have been reconstituted to form the concluding chapter. Thus the reader can benefit from the personal experience, knowledge, and insights of Professor Brooks, whose long-term service in the area of national policy has included membership in the President's Science Advisory Committee, the National Science Board, and the Committee on Science and Public Policy of the National Academy of Sciences. The format employed has allowed me to draw on and to record his own very important initiatives in the promotion of the social sciences in a way that would not have been possible in a chapter of his own composition.

NOTES

[1] The chapters were originally part of a larger work dealing with a number of different aspects of these interactions: I. Bernard Cohen (ed.): *The Natural Sciences and the Social Sciences: Some Critical and Historical Perspectives*, Boston Studies in the Philosophy of Science, vol. 150 (Dordrecht: Kluwer Academic Publishers, 1994). This volume contains chapters by I. Bernard Cohen, Ian Hacking, Victor L. Hilts, Bernard-Pierre Lécuyer, Camille Limoges, Theodore M. Porter, Giuliano Pancaldi, Margaret Schabas, Noel M. Swerdlow, and S. S. Schweber.

[2] The studies by Porter and Wise are of major importance for understanding the interactions between the "exact" sciences (primarily physics and mathematics) with economics in the nineteenth century. In particular, Porter has been exploring some of the aspects of numeracy and quantification in social science at large, while Wise has illuminated the ways in which interactions occurred in the nineteenth century between main-line physics and economics. Richards has been analyzing certain aspects of nineteenth-century social theory, primarily in America and Britain, in its general intellectual-cultural and social background, tracing its roots in the contemporaneous sciences. Schlanger has examined the role of metaphor in organismic theories at large. Stocking has been reorganizing the history of anthropology, showing, inter alia, its contacts with the other social sciences and with certain main aspects of the natural sciences.

[3] Many of their works are cited in various parts of Chapter 1. See Neil de Marchi (ed.): *Non-Natural Social Science: Reflecting on the Enterprise of More Heat than Light*, Supplement to volume 25 of *History of Political Economy* (Durham: Duke University Press, 1993); also Philip

Mirowski (ed.): *Natural Images in Economic Thought: Markets Read in Tooth and Claw* (Cambridge, England: Cambridge University Press, 1994).

[4] Especially their volume *The Invisible Hand: Economic Equlibrium in the History of Science*, translated by Ian McGilvray (Cambridge: The MIT Press, 1990).

[5] See notes 28, 30, and 36 in Chapter 1, infra.

[6] Since my presentation is intended to explore methodological issues, not to produce an encyclopedic survey, I have had to omit many interesting topics, among them the development of anthropology and psychology, two fields for which there is a rapidly growing body of historical scholarship. In this regard, it should be noted that anthropology has had a tradition of writing its history and that psychology has long been known for having produced a large body of distinguished historical writing. See the *Journal of the History of the Behavioral Sciences*. In the case of political science, my only examples (in Chapter 2) are drawn from the seventeenth century, and I take no account here of the vast body of writings on almost all phases of the history of this subject. For similar reasons, I have not discussed the literature concerning history and science.

[7] Although there are few general works on the interactions of the natural and the social sciences, there are many important monographs and articles on particular aspects of this general topic. Some examples to which particular attention may be called are Paul Lazarsfeld: "Notes on the History of Quantification in Sociology," *Isis*, 1961, **52**: 277–333; Bernard Lécuyer and Anthony R. Oberschall: "The Early History of Social Research," *International Encyclopedia of the Social Sciences*, vol.15 (1968), pp. 36–53; A. R. Oberschall (ed.): *The Establishment of Empirical Sociology* (New York: Harper & Row, 1972); and the brief but incisive presentation by Theodore Porter: "Natural Science and Social Theory," pp. 1024–1043 of R. C. Olby, G. N. Cantor, J. R. R. Christie, and M. J. S. Hodge (eds.): *Companion to the History of Modern Science* (London/ New York: Routledge, 1990).

[8] Here, as in the text (notably in Chapter 2), I have somewhat anachronistically referred to social science and to social scientists long before such terms were in actual use. On this subject, see the "Note on 'Social Science' and on 'Natural Science'" in this volume.

[9] On this subject, see the works cited in n. 3 supra and numerous articles in the journal *History of Political Economy*.

[10] 1290^{b}21–1291^{b}13.

[11] *Oxford Classical Dictionary*, 2nd ed., p. 116, §9.

[12] For details and a reproduction of this pseudo-astronomical diagram, see my *Revolution in Science* (Cambridge: Harvard University Press, 1985).

[13] Ibid.

[14] For details, see my "Harrington and Harvey: A Theory of the State Based on the New Physiology," *Journal of the History of Ideas*, 1994, **55**:187–210.

[15] Chapters on these topics by Schweber, Porter, and Limoges are to be found in the volume cited in note 1.

[16] Gini's organismic sociology attained its full expression in his *Il neo-organicismo: Prolusione al corso di sociologia* (Catania: Studio Editoriale Moderno, 1927). Gini also introduced organismic social science in his studies of "economic pathology," as in his *Patologia economica* (Turin: Unione Tipografico-Editrice Torinese, 1923; 5th ed., 1952).

[17] In this theory he concluded that "the upper classes, having low rates of reproduction, will tend to extinction unless they get new members from the lower classes, which have higher rates of reproduction."

ACKNOWLEDGMENTS

Some of the topics presented in this book are derived from my presentations, over many years, in the Seminar on Science, Technology, and Public Policy that was conducted in Harvard's John F. Kennedy School of Government by Don K. Price, Harvey Brooks, and myself.

The research on which this book is based has been generously supported by the Richard Lounsbery Foundation, and I am especially mindful of the courteous consideration and continued kindness of the Director, the late Alan McHenry, whose warm support and friendly encouragement have been helpful during each stage of my research and writing. Alan McHenry saw his role as more than simply providing financial support, and I am especially grateful to him for many suggestions made during our regular meetings and for his incisive comments on an early draft of the present work.

As always, I have a deep gratitude to Julia Budenz, who has worked with me through many drafts of the chapters, helping me to resist the stubborn efforts of each chapter to achieve book-length proportions. I am also thankful that I have been able to call upon Professor Elaine Storella of Framingham State College for research help and for continued assistance in revising and checking my several versions. The research assistance and computer skills of Katharine Appleton Downes have been very important in the completion of this volume.

A special thanks must be given to Robert Cohen, the editor of the Boston Studies in the Philosophy of Science. I am also grateful to Kluwer Academic Publishers and especially Annie Kuypers, the acquisitions editor, for bringing out the volume in which these chapters were originally included. Kluwer Academic Publishers have very kindly given permission to reproduce these chapters here. A good part of the text of the present book has also appeared in an Italian version, under the title *Scienze della natura e scienze sociali* (Rome/Bari: Editori Laterza, 1993).

In dedicating this book to Bob Merton and Tom Kuhn, I not only record an intellectual and personal association of half a century, but take note of the important contribution they both have made to the development of my own ideas and especially of my concept of what science is and does, of the ways in which both the natural sciences and the social sciences advance, and of the relations of ideas to the individuals (singly and collectively) who produce, modify, or otherwise use them. Close friends will appreciate the full significance of the mention of Susan T. Johnson.

Widener Library, Study W
Cambridge, MA 02138

1. AN ANALYSIS OF INTERACTIONS BETWEEN THE NATURAL SCIENCES AND THE SOCIAL SCIENCES

1.1. INTRODUCTION

Ever since the time of Aristotle, the natural sciences and medicine have furnished analogies for studies of governments, classifications of constitutions, and analyses of society.* One of the fruits of the Scientific Revolution was the vision of a social science – a science of government, of individual behavior, and of society – that would take its place among the triumphant sciences, producing its own Newtons and Harveys. The goal was not only to achieve a science with the same foundations of certain knowledge as physics and biology; there was thought to be a commonality of method that would advance the social sciences in the way that had worked so well in the physical and biological sciences. Any such social science, it was assumed, would be based on experiments and critical observations, would become quantitative, and would eventually take the highest form known to the sciences – expression in a sequence of mathematical equations.

By the end of the eighteenth century, it was obvious that no social science had been created as the equal of Newton's physics, Harvey's physiology, or even the new experimental science of electricity pioneered by Benjamin Franklin. On several occasions, Franklin expressed his awareness of this difference between the social sciences (or "moral" sciences) and the recognized physical and biological sciences. In a letter of 1780 to his friend and scientific colleague Joseph Priestley, he took note of the "rapid Progress true Science now makes" and wished that "moral Science were in as fair a way of Improvement." The century's end brought renewed hope for social or moral sciences that would become equal partners with the sciences of nature. A symbol of this dream may be seen in the establishment of the National Institute in France after the Revolution had dissolved the old Royal Academy of Sciences. The new Institute had several "classes," one of which was equivalent in its membership to the old scientific academy, but another was the new "class" of "moral and political sciences" ("classe des sciences morales et politiques"), as a kind of equal partner. Benjamin Franklin had been a "foreign associate" of the old Academy of Sciences since 1773; Thomas Jefferson was elected a "foreign associate" of the new section or class in 1801.

The ultimate fate of this new class of "sciences morales et politiques" is an index of the problems that beset the social sciences. Since social scientists – especially political scientists – cannot help but deal with controversial issues, their opinions and conclusions may become offensive to the ruling powers of the state. Within a very few years of the establishment of the "class," the social and political views of the social scientists in the Institute so ired Napoleon that he reacted by abolishing their class, thus officially severing the ties of social sciences with scientific respectability. The organized physical and biological sciences did not deal with such controversial issues, nor did the group that represented the interests of the members of the old Academy of Inscriptions and "Belles Lettres."

Any historical study of the relations between the social sciences and the physical and biological sciences touches at once on the legitimacy of the several social sciences. A fundamental issue of controversy is whether such legitimacy arises from a slavish adaptation of concepts, principles, theories, and methods from one of the natural sciences (usually considered to be physics) or whether these "other" sciences have their own independent methodologies and standards. In exploring this and allied questions of methodology and legitimacy, our attention will be focused on the late nineteenth century, when two social sciences – economics and sociology – claimed scientific legitimacy because of their use of concepts, principles, and methods of, respectively, physics and biology. An important ground for claiming full membership in the accepted family of "sciences" was a declared general parallelism between these subjects and the accepted sciences of physics and biology, but there was additionally a degree of equivalence of concepts such as energy (utility) or the cell (the social entity of the human individual or the human family). In the case of economics there was even a proud exhibit of equations of identical form with those of physics. We shall see below how these two developments illustrate the two themes of legitimation and of transfer of concept and method.

The present exploration into the impact of the natural sciences on the social sciences leads to several different lines of thought. We shall see that in the late nineteenth and early twentieth century, the physical and biological sciences served two distinct purposes. One was to validate the methodology, the other to guarantee the results. In this enterprise, many founders of the new economics – known today as marginalist or neoclassical economics – chose physics as the science to emulate, but an important school of sociology preferred the biological sciences. The validation of a social science by showing that it is like an accepted natural science may be seen clearly in the example of Léon Walras, one of the late-nineteenth-century founders of the marginalist school of economics. Walras, as we shall see, knew only the most elementary mathematics and very little physics during the decades of the nineteenth century when he was developing his system of economics. It was only later, in the early twentieth century, when he was "hungry" for recognition, that he picked up enough mathematics and physics to claim that his economics was "scientific" and exact because it could produce equations similar in form to those of rational mechanics, the pioneer exact science. Even earlier, William Stanley Jevons had attempted to justify the introduction of the calculus into economics by arguing that this kind of mathematics had been used successfully in rational mechanics – thus implying that economics was like physics because both were susceptible of the same kind of mathematical treatment. Furthermore, Jevons introduced some examples to show that economics, in the form in which he presented it, could be treated like physics, even equating the economics concept of "utility" and the physics concept of "energy."

Validation of a similar kind was sought by those sociologists who adopted the biological sciences as their paradigm. In this emulation, they drew strength from the example of the medical biologist Rudolf Virchow, founder of the doctrine of "cellular pathology," who had introduced social concepts into his medical thought, thus legitimating the association of cell theory and theories of society. Drawing on this association, these sociologists – notably Paul von Lilienfeld, Albert E. Schäffle, Herbert Spencer, and René Worms – constructed a sociology based on such current biological developments as the cell theory, the biological concept of division of labor, medical ideas of normal and pathological, and the physiology of the "milieu intérieure." We shall see that they even introduced little biological tutorials to show the harmony of their ideas with those of the leading biologists of the time.

The marginalist economists differed greatly among themselves with respect to the use of mathematics. The Austrian economist, Karl Menger, for example, did not draw on physics and mathematics. Alfred Marshall, one of the "greats" in this area, preferred a biological to a mathematico-physical model, even though as an undergraduate at Cambridge he had studied mathematics and physics. An important group – including William Stanley Jevons, Léon Walras, Vilfredo Pareto, and Irving Fisher, all of whom claimed that their subject was equivalent to physics – varied greatly in their knowledge of any higher mathematics and the mathematical physics of rational mechanics and energy. Jevons and Walras had, at best, a rudimentary acquaintance with mathematics. Pareto, however, was trained as an engineer and thus, unlike Jevons and Walras, was at home with mathematics and knew some physics. Fisher, who obtained his Ph.D. at Yale, was a student of J. Willard Gibbs and also was qualified as a mathematician. Whereas Pareto and Fisher actually used mathematics in developing their ideas, Walras and Jevons did not, introducing mathematics more as an instrument of legitimation than as a tool of discovery. But the real founder in the application of higher mathematics (i.e., the calculus) to economics was Antoine-Augustin Cournot,[1] who lived somewhat earlier in the nineteenth century and who certainly could not be faulted for his mathematical expertise. We shall have occasion to observe that mathematicians – Henri Poincaré, Henri Laurent, Vito Volterra – criticized the mathematical constructions of the marginalists, challenging the claims that their economics displayed the mathematical integrity of physics.

It is a curious paradox that although the organismic sociologists cannot be censured for their science, their writings seen ridiculous to us today. The marginalist economists are currently under fire for – among other things – not having fully understood the science which they were emulating, yet their ideas are still part of the foundation of today's subject. Furthermore, the kind of physics with which these economists are associated is now outmoded and has been replaced by concepts from relativity and quantum mechanics – subjects that seem not to have permeated deeply (if at all) into today's mainstream economics. Curiously enough, the biological science of the nineteenth century has weathered the years somewhat better than the physics, requiring revisions and expansions but not the same degree of radical restructuring, while the sociology built on the biology has not done as well as the economics which was (in part, at least) linked with the physics. Apparently, the correctness of the emulated science is not intrinsically connected with the permanent value of the resultant social science.

For comparison and evaluation of the different ways in which the natural sciences have influenced the social sciences, at least a rough typology of interactions is required. We shall see that it is sometimes helpful to distinguish between metaphor, on the one hand, and analogy and homology on the other, and also between analogy and homology. The use of metaphor may imply a transfer of values: for example, to demonstrate that economics is a Newtonian science. But analogy implies similarities in function, such as the role of a single great unifying law to explain society in a manner like that of the law of universal gravity in organizing the phenomena of terrestrial and celestial mechanics. Homology, however, implies an identity in form or structure. This similarity, we shall note, may be purely formal. That is, the same equations or principles may appear in two different sciences, which means that there will be an identity of form in which the only differences are the actual letters or symbols in equations or the names of concepts in the statement of principles. In a highly typical example, we shall see that an argument over the economics of "the firm" can be understood by differentiating the use of a general analogy taken from biological evolution and the problems of a specific set of homologies, including such specific concepts as mutation and inheritance. Thus, as we shall note, the distinction between function and form tends to coexist with or modulate into a distinction between the more general and the more specific.

In another example, we shall see why it can be helpful for a critical analyst to make a distinction between analogy and homology in relation to a theory of society. Two major sociologists of the nineteenth century – Paul von Lilienfeld and Albert E. Schäffle – agreed on the importance of using the analogy of the biological cell theory in developing a useful theory of society. They did not, however, reach the same conclusion when it came to the question of specific homology. They parted company on the issue of whether the social homologue of the biological cell was the human individual or the family. In another example that shows why it is important to make a distinction between analogues and homologues, we shall find that Walter Cannon had the laudable idea of applying the results of his research in physiology to social analysis. He wanted to find social analogues of the self-regulating mechanisms which he had been studying in animals and in human beings in his laboratory. So far so good! But he went astray when he sought to introduce specific homologies.

The correctness of the analogues, homologues, and metaphors used by social scientists has never proved to be a guarantee of the validity or usefulness of any social science. Nor is a social science less valid if it does not in any way attempt to imitate a particular natural science, to be like physics or like biology. We shall see, accordingly, that the ultimate criteria for the validity of any social science and the grounds of its usefulness must be independent of the question of whether it is a subject like physics or like biology. Much more important in any evaluation is whether this subject has its own integrity, whether it is internally coherent, whether its results are testable, and whether its assumptions are of the sort demanded by rational explanation. That is, one would not accept a social theory as a science if it depended on the primary postulation of divine intervention. At the same time, of course, a social science which did not take advantage of useful and relevant applications from the natural sciences would be open to severe criticism. Yet this same criticism would be doled out to any branch of the natural sciences that similarly ignored relevant and useful work from other disciplines. Indeed, a development within the social sciences would be equally faulted for ignoring useful and relevant advances in some other of the social sciences. It is the case, however, that a social science like economics – which "looks" somewhat like physics in being quantitative, in finding expression of its principles in mathematical form, and in using the tools of mathematics – tends to rank higher on the scale of both scientists and non-scientists than a social science like sociology or political science which seems less like an "exact science."

1.2. DEFINITIONS AND PROBLEMS

The study of the interrelation between the natural and the social sciences is beset with fundamental difficulties, beginning with the meaning of the two terms: "natural sciences" and "social sciences." Conventionally, the natural sciences comprise the physical and biological sciences, the earth sciences, meteorology, and sometimes mathematics. When I refer without qualification to the natural sciences, I shall be including all of these, from biology and geology to chemistry and physics and mathematics.

The social sciences are generally understood to include anthropology, archeology, economics, history, political science, psychology, and sociology.[2] There is traditionally a third group, the "humanities," embracing such disciplines as philosophy, literary study, linguistic study, and sometimes history. Often the category of science or natural science is extended to some subjects normally regarded as social sciences or as parts of the humanities and may include, in addition to (physical) anthropology and (experimental) psychology, such varied fields as linguistics, archeology, and economics. Sometimes geography is considered a social science, sometimes a natural science. In the last forty years some but not all of the traditional social sciences have come under the umbrella of "behavioral" sciences.[3]

The problem of definition is complicated by the fact that these divisions are not the same in all languages and cultures. Even the designation "science" or "natural science" can give rise to confusion since there are differences in usage among the English "science," the German "Wissenschaft," and the French "science."[4] In English-speaking countries, the term "science" without any qualifying adjective often denotes only the natural sciences considered separate from the social sciences. The Royal Society, the British national "scientific" society, has no membership category for the social sciences[5] and in this respect is even more rigid than its American counterpart, the National Academy of Sciences, which does at present have a recognized category of membership for some social scientists.[6] The French Académie des Sciences is like the Royal Society in excluding social science and is even stricter about admitting non-scientists.[7] In Germany, however, the major academy (the Berlin Academy, founded by Leibniz at the end of the seventeenth century) has always had a broad base of membership.[8] Often in German culture there is a bipartite division of "Wissenschaft" (science or knowledge) into "Naturwissenschaften" (natural sciences) and "Sozialwissenschaften" (social sciences) or into "Naturwissenschaften" and "Geisteswissenschaften" (human sciences).[9]

Furthermore, even within a single language or culture, terms connected with the sciences have not always had the meaning which they bear now. Thus in England as late as the eighteenth and early nineteenth centuries, both "experimental" and "science" were used in a general sense to denote respectively "based on experience" and "system of knowledge,"[10] as well as in senses closer to ours. The older sense of "experimental" and of "science" may be seen in a statement of 1833 by Thomas Babington Macaulay that the "science of government is an experimental science."[11] Macaulay did not mean that this subject was based on laboratory investigations or that it was exactly like physics or biology. For him the science of government was a branch of organized thought, founded on solid experience, especially as revealed by the historical record. A similar use of "experimental" in a political context occurs frequently in the eighteenth century. One example is a letter written by Edmund Burke to the Duke of Bedford, asserting that politics is a "glorious subject" for "experimental philosophy." Another is the subtitle of David Hume's *Treatise of Human Nature* (1739): "an attempt to introduce the experimental method of reasoning into moral subjects." In the introduction to this work, moreover, Hume refers to the "four sciences of *Logic, Morals, Criticism, and Politics*," implying that these subjects are systems of organized knowledge.[12] In Hume's time, the areas of knowledge which we would call science were largely known as "natural philosophy" or "natural knowledge."[13] The term "science" in its present denotation and the associated designation of "scientist" were not introduced into the English language until the nineteenth century and did not become part of general usage until after the 1850s.

For a historian, a striking difference between the natural sciences and the social sciences is the degree to which social scientists still read with profit the classics of their fields, finding an examination of the views of the founders to be instructive and sometimes even necessary for today's subject. So extreme is this practice that James Coleman concludes that university courses in "social theory" today may be regarded as no more than histories of social thought: "An unfriendly critic would say that current practice in social theory consists of chanting old mantras and invoking nineteenth-century theorists."[14] For natural scientists, by contrast, such encounters with the writings of the past are generally held to be unnecessary.

An examination of the literature concerning the relationships between the natural sciences and the social sciences reveals that until fairly recently there was an excessive concentration on whether the social sciences are or are not sciences in the sense of the natural sciences. The experience of many decades has indicated that this is not a fruitful question. Many analysts, such as Hilary Putnam, have insisted that there is no single paradigm which unambiguously applies to all the natural sciences. In most ordinary discourse, the quality of being a "science" is to be like physics. Such an attitude also characterizes the discourse of many scientists – except, of course, naturalists. But even to be like physics has its problems since this category embraces such varied subjects as rational mechanics, experimental optics, and theoretical physics.

There is, in addition, the choice to be made among Newtonian or Einsteinian physics or the physics of quantum mechanics. Probably the one aspect of this question on which most natural scientists would be in agreement is that there is a difference between the natural sciences and the social sciences and that sociology in particular is not a "science" – an opinion held also by certain sociologists.

The question of the social sciences as "sciences" is further complicated by the fact that the answer will depend on the historical period, since the image of what a science is varies from one age to another. Furthermore, one or the other social science may be very like some given natural science and yet be very different from others. Because of the extreme difficulty in setting up hard and fast rules to decide whether a given theory does or does not merit being considered part of "science," we may well understand why, as Robert Merton has indicated, social scientists have allowed this problem "to commit suicide" and have more profitably concentrated on producing "scientific results."[15]

It must be noted, however, that the general problem of definition and delimitation has been of real importance in deciding questions of policy during the past decades. For example, the obvious primary intention of the United States Congress in establishing the National Science Foundation in 1950 was to provide federal support of fundamental research and training in "science," where "science" was intended to signify the traditional natural sciences (including mathematics) and engineering.[16] Many natural scientists at that time were quite vocal in their opposition to the inclusion of any support for the social sciences. For example, the physicist and Nobel laureate, I. I. Rabi, who exerted a very strong influence on questions of science policy, stated bluntly to the Congress, during the debates on the founding of NSF, that government support of the social sciences was inappropriate since it would "strengthen a preconceived point of view or a particular opinion." Additionally, he argued, "most of the things or many of the things which a social scientist has to say are controversial in nature," a feature which – according to Rabi – does not hold for physical science "simply because it is quite objective."[17] Rabi feared that the work done by social scientists, if supported by the new foundation, would reflect adversely on the good work done by natural scientists. Most of the scientific community shared these attitudes.

The hearings with regard to the proposed foundation showed that a significant number of lawmakers opposed support of the social sciences because they tended to equate "social science" with "social reform" and to equate "sociology" with "socialism," a source of confusion that has plagued the social sciences for at least a century.[18] Senator Fulbright, a former university professor, tried to explain to his colleagues that social science is not another word for socialism or "some form of social philosophy."[19] In the event, the attempts to have the social sciences formally incorporated into the National Science Foundation were defeated and the NSF was established by the Congress without any specific provision for the social sciences. A compromise, however, enabled the Director and the National Science Board to exercise full discretion with respect to support of some work in social science within the Foundation, a position that was officially "permissive, not mandatory."

In the inaugural years of the Foundation, the social sciences were all but excluded from aid. Then token support was introduced by an internal administrative decision which permitted direct funding of research in carefully selected areas of the social sciences. First steps in this direction were the extension of the mandate of the biological sciences to include some "behavioral sciences" and the creation in 1955 of a small subdivision of physics (with minimal funding) euphemistically given the neutral designation of "socio-physical sciences." Those of us who were privileged to serve on the inaugural advisory panel of this subdivision represented the history, philosophy, and sociology of science, plus archeology, anthropology, comparative anatomy, political science, sociology and social psychology, and mathematical economics. After some years of steadily increased funds for research, our subjects were incorporated in 1959–1960 into a full-fledged Office of Social Sciences, then reconstituted in 1961 as the Division of Social Sciences, equal in position – though not in prestige, power, or funding – to the other scientific and educational divisions within the Foundation.[20] The National Science Foundation quickly became one of the major sources of funds for research and training in these areas of the social sciences. The existence of this Division, headed by the distinguished sociologist Henry W. Riecken, was a declaration that the social sciences – unlike the humanities – were at last becoming formally and financially recognized as members (although possibly only "associate" members) of the natural sciences establishment.[21]

1.3. TYPES OF INTERACTION

The impact of natural sciences on social sciences involves various components and factors. Among the determinant components are the specification of the area of social science which is to be affected and the choice of the scientific domain which is to provide a source of emulation. These two components are frequently selected together. Another component is the more general one of the scientific climate.

The selection of a particular social science and a particular natural science may be illustrated by a host of examples. In the seventeenth century James Harrington modeled his theory of society on William Harvey's new physiology.[22] Toward the end of the nineteenth century, the economist William Stanley Jevons proposed a new economics based to some degree on the model of Newtonian rational mechanics. In three examples from the last hundred years it was the scientists themselves who designated an area of the social sciences in which their work might be fruitfully applied. The German physical chemist, Wilhelm Ostwald, endeavored to create a new form of social science based on energetics; he called this science "Kulturwissenschaft" instead of the accepted "Sozialwissenschaft."[23] In a somewhat similar fashion the American physiologist Walter Bradford Cannon essayed an extension of his research on self-regulating processes of the human body to social theory, attempting to transform and revitalize the traditional concept of the body politic. In our own time we have seen E. O. Wilson develop sociobiology by generalizing his studies of evolutionary biology and of the group behavior of ants.

A somewhat different example is provided by the British philosopher George Berkeley who – in the eighteenth century – was also working from natural science to social science. He sought to prove that Newtonian rational mechanics might be applied to produce a science of social interactions. This may be likened to the attempt by John Craig, Newton's contemporary, to find a social analogue for law of universal gravity. Adolphe Quetelet, the nineteenth-century pioneer of social statistics, was a professional astronomer who saw in the domain of social numbers a fruitful field for the application of statistical modes of investigation. The opposite path was followed by Emile Durkheim, who discerned in the social numerical data of suicides a statistical base for a science of society.[24]

The more general determinant component, the scientific climate, may be observed in almost every instance in which there is an impact of the natural sciences on the social sciences. In the seventeenth century, the creation of new mathematics – analytic geometry, the calculus, and the use of continued fractions and infinite series – and the outstanding success of a mathematical point of view in physics and astronomy established a mathematical climate, the effects of which are easy to see in the social sciences.[25] Hugo Grotius, whose intellectual ideal was Galileo's new physics of motion, displayed the influence of the mathematical way of thinking in his celebrated treatise on international law. In this climate the French engineer Vauban saw the need for a number-based statecraft. Perhaps the most easily discerned effect of this mathematical climate is the development by Graunt and Petty and their eighteenth-century successors of a numerical approach to the problems of government which Petty named political arithmetic.[26]

In the late eighteenth century, the scientific climate was in some respects even more mathematical. In this era mathematics had two different implications for the natural sciences: to apply actual mathematical procedures in order to derive principles of science from sound axioms and to base science on numbers or on quantitative considerations. Even natural history, that least mathematical subject within the natural sciences, began to incorporate some quantitative features, as we may see in Buffon's celebrated *Histoire naturelle*, where the discussions of anthropology featured the statistical studies of mortality made by Jean-Pierre Emile Dupré de Saint-Maur.[27] The development of a mature science of probabilities, remarkably advanced by Laplace's *Théorie analytique des probabilités* of 1812, was another very significant aspect of the quantitative scientific climate at this time. It had a notable counterpart, of course, in the collection of all sorts of demographic data and social statistics.[28]

The influence of the mathematical climate may be seen also in the concept of an "ideal man," Condorcet's primitive earnest of Quetelet's later concept of "l'homme moyen" or "average man." Condorcet's model for social science, as Keith Baker has shown,[29] embodied the new probabilistic philosophy which made this area of knowledge as susceptible of calculation as the physical sciences, a fundamental step in a sequence that eventually led to Quetelet's statistically based "physique sociale."[30] Baker argues that "the structure of scientific discourse in the eighteenth century not only yielded a probabilistic model of science explicitly applicable to social affairs but, in some ways, required such application as proof of the validity of scientific knowledge."[31]

The intellectual climate, furthermore, includes the standards of knowledge and a system of values that constitute a set of metaphors which determine a style of doing science acceptable to the members of the profession. The introduction of mathematical methods from physics into economics shows how the choice of a metaphor from the natural sciences may condition such acceptability. The "progenitors of neoclassical economic theory" of the latter nineteenth century and early years of the twentieth wished "to preserve the legacy of the classics and to refurbish their thoughts in line with new ideas," and therefore they "boldly copied the reigning physical theories." These words sum up the controversial findings of Philip Mirowski, who has been exploring the interplay of mathematical physics and economic theory in the 1870s and later. These "neoclassicals," Mirowski claims, "did not imitate physics in a desultory or superficial manner," but rather "copied their models mostly term for term and symbol for symbol and said so."[32] The economists Jevons, Pareto, and Fisher declared a goal of making economics a "true" science, choosing physics as a model because this was the science they knew best and because physics was esteemed for its intellectual success and was characterized by the extensive use of mathematics, the primary feature which these economists regarded as making a subject scientific.[33] In this process we not only see the pressure of the intellectual or scientific climate determining a model of mathematics and physics for economics but also discern the value-laden aspects of the particular scientific model chosen by social scientists. The economists apparently did not favor mathematical physics, primarily energy physics and rational mechanics, only because this part of the natural sciences seemed to offer the most fruitful source of useful applications; rather, they opted to emulate a part of the exact sciences that had the highest standing and that could thereby confer upon their own endeavors the quality of legitimacy, showing that their subject exhibited the features of an exact science.[34]

It must not be concluded, however, that even in so physics-like a discipline as economics, the introduction of the techniques and discourse of the exact sciences was an easily acceptable means of showing that one's work was "scientific" or of winning the respect of fellow members of the economics community. Writing about his own experience in introducing the techniques of mathematical thermodynamics into economics, Paul Samuelson has observed that his critics supposed that he was attempting "to inflate the scientific validity of economics," even "perhaps to snow the hoi polloi of economists who naturally can't judge the intricacies of physics." Not so! "Actually," he goes on, "such mathematical excursions, if anything, put a tax on a reputation rather than enhancing it."[35] He had to overcome the impression of being a brash young man and of flouting the agreed-upon rhetoric, metaphors, and standards of technical discourse of his profession.

In addition to the determinant components which we have been considering, the impact of natural sciences on social sciences involves various qualifying factors. These include the degree to which the state of the chosen part of social science permits the desired input from the natural sciences, the degree to which the developments in the natural sciences are susceptible of such application, and the justness of the fit. With regard to whether the chosen part of social science permits the desired input from the natural sciences, an example is once again provided by political arithmetic. Laudable as was the aim of Graunt and Petty to reduce questions of polity to mathematical considerations, the numerical demographic data were not adequate for the purpose and hence did not permit the desired application. By contrast the subject of economics in the mid-nineteenth century proved to be well adapted to the application of mathematical techniques, as may be seen in the successful construction of mathematically based theories by such economists as Edgeworth, Jevons, and Walras.

Whether or not the chosen part of social science is suitable for the application of a particular input often involves the state of development reached by a subject at a given time. One reason why Quetelet had greater success in creating a statistically based social science than Petty or the eighteenth-century political arithmeticians was that in the nineteenth century the actual raw materials of social science – the demographic, census, and social data – were more abundant and reliable than in the eighteenth.[36] Of course, there was the additional factor of the creation of modern statistical methods – in part by social scientist themselves – during the late eighteenth and early nineteenth centuries. Both of these causes for Quetelet's success and Petty's failure are part of the conditioning factor which consists in the state of development of the social science involved.

A second qualifying factor is of the opposite sort from the first one: it is the degree to which the natural sciences have developed to a state that will permit the desired application. The example of political arithmetic exhibits this factor because neither arithmetic nor elementary algebra is sufficient for the analysis of demographic or social data. There was need for a new mathematics, the mathematics of probability, that could be applied to statistical data. The nineteenth-century social scientists who sought to create a numerically based science of society did not wait for a suitable model of applied statistics to emerge from the physical or biological sciences. Rather, since Quetelet and others recognized that the mathematical techniques of statistics had developed sufficiently to permit a wide range of applications, they moved ahead on their own to create a statistically based social science. The high level of statistical social science which they produced then served as a model for emulation in the exact sciences – in the physics of Maxwell and of Boltzmann.

These two qualifying factors partake of opposite facets of the justness of the fit. Of major significance here is the degree of exactness of analogy between some part of social science and some primary concepts from the natural sciences, a topic further explored in the following sections. Or it may be that the structure of some part of the social sciences (for example, economics) may have such a strong formal resemblance to some aspect of the natural sciences (say rational mechanics) that similar equations, laws, and principles may apply to both. This is a familiar situation within the natural sciences; for example, the equations for an alternating current proved to be formally identical to those for an oscillating pendulum. The late nineteenth century witnessed such a fit between a generalized concept of evolution, developed in the context of biological science, and the study of societies or cultures. Many instances of both close and poor fit prove to have two very different aspects, which may be termed analogy and homology.

1.4. ANALOGY AND HOMOLOGY

In considering the interactions of the natural sciences and the social sciences, a useful distinction may be made between analogy and homology and between both of these and metaphor. The word "analogy" is generally used today to indicate many kinds of similarity, but in the natural sciences analogy denotes an equivalence or likeness of functions or of relations or of properties. Thus David Brewster wrote in

1833 about waves or undulations as "a property of sound which has its analogy also in light."[37]

This particular sense of analogy is of special significance in writings on natural history: for example, to express a similarity in function between organs which may seem somewhat different in different species. An example is the wing of a bird as compared with the wing of a bat. Wings of each type enable their possessors to fly, and hence they are analogues; that is, they perform similar functions in both animals, even though a bird's wing is covered with feathers while a bat's wing is a stretched skin membrane.

In the language of the life sciences, the term "homology" has a specific meaning which is quite distinct from that of analogy: to denote similarity in form as distinguished from similarity in function.[38] The distinction becomes apparent once attention is focussed on structure (anatomical construction) rather than function (use in an action).[39] An anatomical comparison of bone-structure shows that the wings of the bat resemble the wings of birds, the forelegs of quadrupeds, and the arms of humans. Hence, the wing of a bird and of a bat, the foreleg of a quadruped, and the arm of a human (and also the pectoral fin of a fish and the flipper of a seal) are homologues. It should be noted that in evolutionary biological science,[40] "homologous" has a strict signification: a correspondence in the type of structure of parts or organs of different organisms resulting from their descent from some common remote ancestor.[41]

In what follows I shall consider the terms analogy and homology as denoting respectively, at their most precise, similarity in function and similarity in form. But the differences between these two kinds of resemblance may result, as will be shown, in a related and sometimes more obvious difference between analogy as suggesting only a general similarity and homology as representing a quite specific one. These distinctions will help to indicate the ways in which the social sciences have used the natural sciences and equally the ways in which the natural sciences have used the social sciences. The same features may be seen in the ways in which the different natural sciences have made use of one another.[42]

Several examples of laws formulated for the social sciences illustrate the distinction between analogues and homologues. A number of social laws in the domains of human behavior, sociology, and economics were proposed as either analogues or homologues of the Newtonian law of universal gravity. The Newtonian law accounted for a number of different kinds of phenomena both in the heavens and on our earth. These phenomena included the orbital motions of planets, planetary satellites, and comets; the occurrence of the tides in the ocean; the fact that, at any given place, bodies of different weights fall at the same rate; the varying of terrestrial weight with latitude; and much else. The Newtonian law states that the force of gravity between any two bodies is directly proportional to the product of the masses of the bodies and inversely proportional to the square of the distance between them.

In the middle of the nineteenth century the French economist Léon Walras and the American economist and sociologist Henry C. Carey proposed laws which can be considered analogues of Newton's to the degree to which both were intended to serve the same basic function in sociology or economics that Newton's law served in rational mechanics and celestial dynamics. Carey's law was presented as a kind of corollary to a general principle of social gravitation: "Man tends of necessity to gravitate towards his fellow-man." His corollary is that "the greater the number [of men] collected in a given space the greater is the attractive force there exerted."[43] Like Newton's law, Carey's expresses a property of an "attractive force." Carey's force is as the number of men in two places, which is formally equivalent to Newton's force as directly proportional to two masses. That is, a force is posited as proportional to a product of two variables; in this sense there is a homology between the two laws. In Carey's law, however, the force is inversely as the distance, whereas in Newton's law the force is inversely as the square of the distance.[44] The two laws, therefore, do not really have the same form; there is not a perfect fit. This kind of failure in homology may be considered an example of mismatched homology, in a sense somewhat analogous to Alfred North Whitehead's concept of the fallacy of misplaced concreteness.[45]

Furthermore, in Carey's law the number of men is an unsatisfactory homologue of Newtonian mass. Mass is the characteristic concept of Newtonian or classical physics and was invented by Newton. Newtonian mass is an invariant property of any body or sample of matter; it does not change when the body is heated or chilled, bent or twisted, stretched or compressed, or transplanted to another location, whether this is another spot on earth or some place out in space or even on the moon or on another planet. In this feature it differs from a local property such as weight, which varies with latitude on earth and also with transplantation to the moon or to another planet.[46] Although Carey's concept fails as a homologue of Newton's mass, it has the same function in his law that Newton's concept has in his law of universal gravity, that is, it shows society functioning in a way that is similar to the way in which Newton shows matter functioning. In short, the two concepts are used analogously even though they are not homologous. But the specificity of comparison and, in particular, the patent attempt to assert a similarity of form between his law and Newton's compel us to characterize Carey's laws as involving an unsuccessful homology.

Let me now turn to Walras's law. Early in his career, in 1860, Walras wrote a short work on "The Application of Mathematics to Political Economy." Here he essayed a Newtonian law of economics, that "the price of things is in inverse ratio to the quantity offered and in direct ratio to the quantity demanded."[47] This law may be considered an analogue of the Newtonian law of gravity in the sense that it is supposed to have the same important role in market theory that the Newtonian law has for the theory of planetary motion; that is, it displays a functional relation between economic entities that has the same functional role as Newton's. But while the two laws may be regarded as analogues in the sense of being functionally equivalent, and even though Walras's law is presented in a form much like Newton's, Walras's law and Newton's are not genuinely homologous. First of all, Walras's law depends on a simple inverse ratio (the price is inversely proportional to the quantity offered), whereas Newton's law invokes the ratio of the inverse square (the force is inversely proportional to the square of the distance). Second, Walras's law involves a direct proportion of a single quantity or parameter (quantity demanded), whereas Newton's law uses the direct proportion of two quantities (the masses). Furthermore, Walras's law posits a price that is proportional to a "quantity" divided by another "quantity" of the same kind or dimensionality, that is, proportional to a dimensionless quotient or pure numerical ratio. Clearly, whatever other characteristics this law may have, it exemplifies a mismatched homology.[48]

The Newtonian social laws of Carey and Walras may be contrasted with Berkeley's attempt to produce a social science based on gravitation. In terms of my earlier discussion of determinant components, I may take note that Berkeley's point of departure was natural science whereas that of Carey and Walras was social science. Moreover, unlike Carey and Walras, Berkeley was an astute student of Newton.[49] Writing in 1713, he began by stating the principles of Newtonian celestial dynamics correctly. This was no mean feat since many social scientists of the eighteenth century, such as Montesquieu,[50] held a totally incorrect view of Newtonian celestial physics. They believed that planets and other orbiting bodies are in a state of equilibrium,[51] a supposed balance between a centripetal and a centrifugal force.[52] Berkeley asserted[53] that society is an analogue (a "parallel case") of the Newtonian material universe and that there is a "principle of attraction" in the "Spirits or Minds of men."[54] This social force of gravitation tends to draw men together into "communities, clubs, families, friendships, and all the various species of society." Furthermore, just as in physical bodies of equal mass "the attraction is strongest between those [bodies] which are placed nearest to each other," so with respect to "the minds of men" – *ceteris paribus* – the "attraction is strongest . . . between those which are most nearly related." He drew from his analogy a number of conclusions about individuals and society, ranging from the love of parents for their children to a concern of one nation for the affairs of another, and of each generation for future ones. Although Berkeley introduced the notion of social attraction and regarded the "minds of men" and the closeness of their relation as having social roles similar to those of mass and distance, he did not attempt to develop an exact homology of concept, nor did he quantify his law of moral force. Perhaps he was thereby spared any possible mismatched homology.[55]

David Hume's *Treatise of Human Nature* (1738) provides an example, similar to Berkeley's, in which there is a general analogue of the Newtonian law of universal gravity without any proposed homology. Hume's goal was to produce a new science of individual human moral behavior that would be equivalent to Newton's natural philosophy.[56] He stated that he had discovered in the psychological principle of "association" a "kind of ATTRACTION, which in the mental world will be found to have as extraordinary effects as in the natural, and to show itself in as many and as various forms."[57] In short, he believed that psychological phenomena exhibit aspects of mutual attraction. But he did not propose a law of mental gravity as a direct counterpart to Newton's law, nor did he propose concepts homologous to those of Newton's *Principia*.[58]

The foregoing examples, in addition to illustrating aspects of analogy and homology, indicate how the natural sciences have influenced the social sciences. In each case there was an attempt to create a Newtonian social science by introducing concepts or laws intended to be analogues or homologues of those used by Newton in his rational mechanics. But whereas Carey and Walras may be characterized as having formulated unsuccessful homologues, Berkeley and Hume may be regarded as having presented only analogues. And there were other social scientists in the eighteenth century and in the nineteenth whose expressed goal was the less specific one: to create a social science that would somehow be the equal of Newton's system only to the extent of organizing the phenomena of society in the same manner in which Newtonian science had organized the phenomena of the physical and cosmic realms.

An outstanding example of such an attempt to produce an analogue of Newtonian science without any homologues occurred in the early nineteenth century in the system of Charles Fourier. Fourier claimed to have discovered an equivalent of the gravitational law, one that applied to human nature and social behavior. Likening his discovery to Newton's, Fourier even alleged that he had been led to his discovery by an apple. He boasted that his own "calculus of attraction" was part of his discovery of "the laws of universal motion missed by Newton."[59]

When, in 1803, Fourier announced his discovery of a "calculus of harmony," he declared that his "mathematical theory" was superior to Newton's, since Newton and other scientists and philosophers had found only "the laws of physical motion," whereas he had discovered "the laws of social motion." Fourier's social physics was based on a system of twelve human passions and a fundamental law of "passional attraction" or "passionate attraction," from which he concluded that only a carefully determined number of individuals could live together in "harmony" in what he called a "phalanx."[60] This Newtonianism was based on a very general Newtonian analogy and contained no homologues of concepts or laws from Newtonian physics.

Emile Durkheim provides another example of a claim to have discovered a social analogue of Newton's law of universal gravity. This emulation of Newtonian physics is all the more surprising in that it appears toward the conclusion of Durkheim's *Division of Labor in Society*, a work exhibiting extensive use of organismic – i.e., biological and medical – analogues of society, even introducing biological cells, physiological functions, the action of a nervous system, and other anatomical and morphological elements. Durkheim's Newtonian social law depends on two social factors: "the number of individuals in relation ["en rapport"] and their material and moral proximity." These factors likewise, are for him, "the volume and density of society"; their increase produces the "intensification which constitutes civilization," or, as he expresses the same idea in a note, "growth in social mass and density" is "the fact which determines the progress of the division of labor and civilization." Durkheim proudly called the sociological law which he had discovered the "law of gravitation in the social world."[61] And one of his formulations of this law certainly is an echo of Newton: "*The division of labor varies in direct ratio with the volume and density of societies, and, if it progresses in a continuous manner in the course of social development, it is because societies become regularly denser and generally more voluminous.*"[62]

Durkheim's law states that "all condensation of the social mass, especially if it is accompanied by an increase in population, necessarily determines advances in the division of labor."[63] That is, in his terms, any increase in social volume or density must result in a heightened competition among similar occupational groups, which will produce a greater division of labor or occupational specialization.[64] Durkheim did not offer evidence of detailed numerical data to support his Newtonian law, nor did he ground it in principles of physics. Rather, he justified the law primarily by means of a biological analogy, a law of Darwin's.[65]

Durkheim's "law of gravitation in the social world" partially resembles Newton's law, since it invokes concepts similar to Newtonian mass, volume, and density. Nevertheless, Durkheim's law does not deal in a Newtonian manner with the interaction of two groups or societies, or with the factor of the distance between the elements of such a pair. He was presumably implying no more than an analogy between the fundamental character of his social law of gravitation and Newton's physical law. He asserted the importance of his discovery of "the principal cause of the progress of the division of labor" by declaring that it has revealed "the essential factor of what is called civilization."[66]

The examples of Durkheim and Fourier, like those of Hume and Berkeley, exhibit a significant feature of the distinction between analogy and homology. Analogies may be useful or useless, appropriate or inappropriate, and moderate or extravagant, and they can be evaluated for their relevance. Homologies, by contrast, are subject to evaluations in terms of correctness rather than relevance, since they imply an identity of form or structure. Carey and Walras proposed laws that were meant to be Newtonian, but that – by objective standards – did not match the original. They were also so specific that they entailed homologies which can be judged whether they were closely matched. Berkeley and Hume were content with rather general analogies and therefore cannot be faulted on grounds applicable to Carey and Walras. And it is the same for Fourier and Durkheim.

Although errors in homology do not occur in Fourier's and Durkheim's sociologies, mismatched homology characterizes another current of nineteenth-century social thought and its twentieth-century overtones, the attempts to produce organismic theories of society. Examples may be found in the writings of such diverse authors as Thomas Carlyle, Johann Caspar Bluntschli, Paul von Lilienfeld, Albert E. Schäffle, René Worms, A. Lawrence Lowell, Theodore Roosevelt, Herbert Spencer, and Walter B. Cannon.[67]

Mismatched homology appears as a prominent feature in Thomas Carlyle's analysis of the problem of society in *Sartor Resartus* (1833–1834). An example is provided by his discussion of the social analogy of the skin:

> For if Government is, so to speak, the outward SKIN of the Body Politic, holding the whole together and protecting it; and all your Craft-Guilds, and Associations for Industry, of hand or of head, are the Fleshly Clothes, the muscular and osseous Tissues (lying *under* such SKIN), whereby Society stands and works; – then is Religion the inmost Pericardial and Nervous Tissue, which ministers Life and warm Circulation to the whole. Without which Perocardial Tissue and the Bones and Muscles (of Industry) were inert, or animated only by a Galvanic vitality; the SKIN would become a shrivelled pelt, or fast-rotting raw-hide; and Society itself a dead carcass, – deserving to be buried.[68]

Carlyle appears to have been obsessed with such organismic comparisons drawn from the realms of anatomy and medicine. For him, England was "in sick discontent," writhing "powerless on its fever bed," and the evils of his contemporary world were a kind of "Social Gangrene."[69]

Another nineteenth-century social thinker who was obsessed with extravagant organismic comparisons was Johann Caspar Bluntschli, a Swiss-German jurist who spent a number of years as a professor at Heidelberg.[70] He was author of many books on the state and on society, but his major theoretical work was *The Theory of the State* (1851–1852; 6th ed., 1885–1886), and his most extreme work was his *Psychological Investigations concerning State and Church* (1844).[71] Deeply influenced by the mystic-psychologist Friedrich Rohmer,[72] Bluntschli endowed the state with the sixteen psychological functions that he believed characterized human beings.[73] Convinced that both the state and the church are organisms similar to human beings, Bluntschli quite logically concluded that they both must have all the primary human attributes, including sexual characteristics, the state representing "the male, the church the female element." This attribution of sex led him to a theory of history, based on social-sexual development, in which the historical "evolution" of society and the state followed the pattern of "evolution" of single individuals. Tracing the sexual history of church and state from childhood (the ancient Asiatic empires) through adolescence (the Jews of Biblical times) to early maturity (classical Greece), he found that in Greece[74] "the ecclesiastical organization" matured earlier "than the political institution," just as "the girl ripens earlier than the boy." So extreme is Bluntschli's mismatched homology that a reader may find it difficult to imagine that he was developing a social parallel when he went on: "The sexual organs of the girl are sooner developed than those of the boy. The youthful breasts begin to swell; and the unfolding virgin turns into a beauty. Beauty was the soul of the cult of the Hellenes. . . ."[75] Bluntschli's attitude towards the sexes led him to assert that the papal desire to subordinate the state to the church is as "unnatural" as "the subordination of a husband to his wife in a household." He envisaged a time, not far off, when the "male state will reach full selfhood," when the "two great powers of humanity, state and church, will appreciate and love each other, and the august marriage of the two will take place."[76]

A similar extravagance occurs in the organismic conception of society proposed by the Russian sociologist, Paul von Lilienfeld, in the comparison which he made between the intellectual and moral state of a hysterical woman and a condition of society.[77] As the physiological foundation of this likeness, he used in particular the findings reported by Dr. Edmond Dupouy (ca. 1845–1920), author of numerous works on medicine, psychology, and medical history. Quoting Dr. Dupouy, Lilienfeld described the condition of women suffering from hysteria.[78] They are, he noted, "mobile in their sentiments," and "they pass very easily from tears to laughter, from excessive joy to sadness, from passionate tenderness to haughty rage, from chastity to wanton purposes and lewd ideas." Additionally such women "love publicity, and to get themselves talked about they employ every means: denunciation, simulation of infirmities or sicknesses, and the revolver." They find joy in pretending to be "victims of anything; they say they have been violated." In order to "achieve their goals they deceive everyone: husband, family, confessor, examining magistrate, and their doctor."[79]

The reader who is uninitiated in the literature of organismic sociology may wonder what social manifestations could possibly be the counterparts of these symptoms. Lilienfeld develops the comparison by presenting a series of correspondences which clearly must be characterized as homologues even if he refers to analogy. He begins by asking rhetorically whether the symptomatic behavior of women suffering from hysteria is not "perfectly analogous to the manner in which the population of a large city behaves during a financial crisis or on the occasion of civil disturbances." He finds in the behavior of such women "a faithful picture of the agitation of parties during elections." And when we consider the past, he asks, do we not find the same confused and disordered pattern of behavior, "caused by convulsive and contradictory reflexes of the social nervous system," during "all the religious, economic, and political revolutions with which humanity has been assailed?"[80] This complex nesting of mismatched homologies needs no comment.

Two authors of very different sorts, one from the nineteenth and one from the twentieth century, provide additional case histories that illustrate the easy susceptibility of social thought to mismatched homology. The first, Herbert Spencer, was a self-educated sociologist and philosopher; the second, Walter Bradford Cannon, was an eminent scientist who dabbled in sociology.

Herbert Spencer[81] indulged himself in analogies and homologies. An extreme example of mismatched homology, which even his sympathetic biographer must admit is a case of "dubious biology . . . added to pedestrian sociology," is Spencer's likening of "the coalescence of the Anglo-Saxon kingdoms into England" and the formation of crustaceans.[82] Here he was introducing his own odd notion that crustaceans, like insects, are "composite animals," in which the segments are independent life-units joined together.[83]

Although he also drew on parallels from the physical sciences, organic correlations permeate Herbert Spencer's writings on sociology.[84] Two samples of his extremes in the production of homologues are (1) his comparison of "the undifferentiated and fragmented structures of Bushmen" with "the protozoa" and (2) his likening of "the ruling class, the trading or distributive classes, and the masses" to "the mucous, vascular and serous systems of the liver-fluke."[85] Perhaps the limit is reached when he refers to the two great national schools of France as "a double gland" intended "to secrete engineering faculty for public use."[86] This final Spencerian example is comparable to one introduced by René Worms in the early twentieth century, and based on regeneration in marine animals such as starfish. Citing the authority of Spencer, Worms compared the way in which certain animals replace a destroyed or damaged organ with Chancellor Maupou's dismissal of the Parlement of Paris and its replacement by a new assembly.[87]

The example of Walter Cannon is more interesting than that of Herbert Spencer because Cannon was one of the foremost scientific investigators of his time. His first essay in biological sociology (1932) was titled "Relations of Biological and Social Homeostasis,"[88] an exploration of whether equivalents of the "stabilizing processes" in animal organisms can be found in "other forms of organization – industrial, domestic or social." In a manner reminiscent of Spencer and other nineteenth-century organicists, Cannon compared the circumstances of small groups of humans living in "primitive conditions" to the "life of isolated single cells," and the grouping of "human beings . . . in large aggregations" to cells "grouped to form organisms."[89] Only in highly developed organisms, he reported, do the "automatic processes of stabilization" work "promptly and effectively." The comparison seemed to show that our present social system resembles organisms low on the evolutionary scale or organisms that have not fully developed, in both of which "the physiological devices which preserve homeostasis are at first not fully developed."

Cannon's major field of scientific investigation was the study of self-regulating processes in the human (and animal) body, stressing the role of the "milieu intérieure." Accordingly, his announced goal in studying social systems was to find in "a state or nation" an "equivalent" for the "fluid matrix of animal organisms." And it is here, in the suggestion of an analogy, that Cannon reveals the naive quality of his social thought. In the social body, he wrote, the equivalent ("in a functional sense") of the fluid matrix for maintaining homeostasis in the living body, is

> the system of distribution in all its aspects – canals, rivers, roads and railroads, with boats, trucks and trains, serving, like the blood and lymph, as common carriers [on which] the products of farm and factory, of mine and forest, are borne to and fro.[90]

Although Cannon sought to limit his comparisons to functional analogies, he unwittingly fell into the trap of mismatched homology by making his analogies far too substantive. He simply could not restrain himself from introducing homologies when he was comparing the cells in an organism with the members of a social group, or the lymph and blood with the system of canals, rivers, roads, and railroads.

Cannon's essay illustrates the danger of using apparent likenesses. On the level of general analogy, his suggestion that society resembles an organism could be regarded as original and instructive, at least by implying that the stability of a society is caused by certain self-regulating mechanisms. We may agree with Robert Merton, however, that Cannon made the mistake of introducing "substantive analogies and homologies between biological organisms and social systems." Merton went so far as to describe Cannon's result an "unexcelled . . . example of the fruitless extremes to which even a distinguished mind is driven." This comment is all the more significant in that it occurs in Merton's essay on "Manifest and Latent Functions,"[91] in which he finds "Cannon's logic of procedure in physiology" to be a model for the sociological investigator, recommending that his readers study Cannon's book on the *Wisdom of the Body*, while warning them about "the unhappy epilogue on social homeostasis."

Almost ten years later Cannon returned to this topic, choosing it as the subject of his presidential address to the American Association for the Advancement of Science, delivered in December 1941.[92] In preparing the new version, Cannon sought help and advice from a sociologist, his junior colleague Robert K. Merton, who sent him a list of books and articles on the subject of society as an organism. Cannon now withdrew his earlier assertions about similarities between cells and human members of society, and he declared that comparisons of "the body physiologic and the body politic" had been discredited in the past because they had mistakenly concentrated on "minutiae of structure."[93] He came out strongly against what he considered to be absurd (we would say "mismatched") homologies. We are "not illuminated," he said, "by a likening of manual laborers to muscle cells, manufacturers to gland cells, bankers to fat cells, and policemen to white corpuscles." He, accordingly, would not be concerned with structures but would rather examine "functional accomplishments in physiological and social realms." Yet, when he posed once again the earlier question of what "corresponds in a nation to the internal environment of the body," his reply was essentially the same as before: "The closest analogue appears to be the whole intricate system of production and distribution of merchandise."

In his presentation of the nation's equivalent of the body's fluid matrix,

Cannon now omitted canals and boats (although he kept the rivers) and added "all the factors, human and mechanical, which produce and distribute goods in the vast and ramifying circulatory system which serves for economic exchange." In less florid prose than before, he said: "Into this moving stream, products of farms and factories, of mines and forests, are placed at their sources, for carriage to other localities." His own display of substantive analogies or mismatched homologies was as unfortunate as it had been in the earlier presentation. As the lawyers say, *Res ipsa loquitur*.

In considering these examples of mismatched homology, our evaluations may be sharpened by attention to the reasons why they seem outré to a critical reader. Why do we smile and assume a condescending air when we read the writings of organicist sociologists like Bluntschli, Lilienfeld, and Spencer, but not when we encounter physical models such as Jevons's lever or Walras's economic machine, both of which will be discussed below, or the numerous attempts to find in the realms of social sciences an analogue of the Newtonian universe? The reason is not simply that one set is biologically based while the other set comes from physics. Henry Carey's attempt to produce a sociology based on electricity, a later rival to his astro-sociology, may provoke our smiles and giggles just as easily as the systems of the organicists.[94]

I believe that our pejorative evaluation of certain social comparisons is based at least partly on the fact that the biological equivalent is usually a real object, an actual living being, endowed with all the forces of life and subject to all of life's problems, such as disease, aging, anxieties. By contrast, the parallels from physics are not concrete but abstract and theoretical. Jevons's lever is actually a mathematical lever and thus does not have such material properties as color, hardness, weight, or physical dimensions other than length. Correlations based on a gravitational universe make use of abstract concepts, just as Newton did in Book One of the *Principia*.[95] That is, in Book One there are no real planets with material sizes, shapes, and similar properties but only mass points whose properties are position in a mathematical space, mass, and the power to give rise to, and to be acted on by, a gravitating force. Thus, unlike the earthy biological sources of comparison, those from physics tend to be abstract[96] and may even serve primarily as sources of equations.[97]

Where Bluntschli, Lilienfeld, and Spencer argue that society is itself an organism or is very much like an organism, the "mechanical economists" – Stanley Jevons, Léon Walras, Vilfredo Pareto, Irving Fisher – declare that economics is analogous with mechanics because of the close similarities between the equations of economics and those that originate in classical mechanics. The problem with the organicists' conception of society is, therefore, not that they found their parallels in living systems but that they did not place their considerations on a plane of abstraction, as did those who drew on analogies from physics. They were extravagant because their goal was to create a homology rather than a general analogy. Their procedure is very much like what Whitehead described as "the accidental error of mistaking the abstract for the concrete," mistaking the abstractions of social theory for the concreteness of an actual biological organism. It is not an error to make use of organicist analogies (or in figures which we shall study below as metaphors) in discussing society at large, the political system, or the economic system. People constantly use expressions deriving from the organic notion of the body politic, such as head of state, nerves of government, healthy state of society or of the economy, consumption, arteries, and many others. Werner Stark, one of the severest critics of organicist theories of society, who describes Lilienfeld's theories as "ravings" and "nonsense," nevertheless admits that in writing about certain aspects of society

> one is constantly tempted to express them in organismic similes: phrases like 'one sector limps behind' or 'one sector is out of joint with the rest' tend to form themselves, as of their own volition, in one's mind, and try to push themselves into, and to flow out of, one's pen. This alone shows that organicism has a deep root, and that its basic metaphor is not absurd, even if its votaries make it so.[98]

Much of today's discourse on society, social problems, or systems of social thought, and on political systems and the state, continues to make use of images related to living systems even though usually there are no longer any of the extremes that characterized the nineteenth-century organicists and some of their early twentieth-century successors. Current usage tends to be on the level of analogy and metaphor and not of homology, making use of the general and the abstract rather than the specific and the concrete.[99]

1.5. METAPHOR

Thus far we have been considering analogies and homologies, but we have not yet addressed the general problem of metaphor.[100] When we treat metaphor in relation to the interactions between the natural and the social sciences, it is sometimes useful to make a distinction among four levels of discourse involving comparison. One extreme level is metaphor, the other is identity, with analogy and homology as intermediary. These four levels of discourse may be easily illustrated by reference to biology and physics as utilized in the social sciences.

First, identity. "What is a society?" asked Herbert Spencer; his reply was, "An organism."[101] Two others of the "identity" persuasion were Otto Bluntschli, who, as we have seen, endowed society and its institutions with sex, and Paul von Lilienfeld, who, as we shall see, declared explicitly – for example, in the title of one of his major works – that he considered society to be a "real" organism. Also to be placed in this category are Albert Schäffle, despite some qualifications which he made in theory, and René Worms, at least in his earlier phase. Those whose belief was at the other extreme merely wrote figuratively of society as generally like an organism or as like an organism in some specific respects; they adopted an organismic metaphor. Their number includes Emile Durkheim, Walter B. Cannon, and René Worms in his later works. The level of metaphor has been a consistent feature of the concept of the body politic, which has successively illustrated the changes in physiology and medicine, being Galenic until the seventeenth century, then Harveyan, and so on.[102]

Traditionally, a metaphor is a literary figure of speech, aesthetic or rhetorical. For Aristotle, a metaphor gives something a name that properly belongs to something else.[103] Because metaphor and analogy both invoke features of similarity as well as contrast, it is easy to understand why a clear distinction is not always made between them. Historically, these two were closely linked; Aristotle held that an analogy is only a special case of metaphor.[104] Furthermore, even a specificity akin to that of homology may be regarded as metaphor if the usage is primarily literary – that is, aesthetic or rhetorical – rather than being chiefly an aspect of logical argument.

Metaphor has long been used as a rhetorical device to enhance oral and written communication so as to increase the effectiveness of the message delivered, but during the Scientific Revolution of the seventeenth century rhetoric fell into disfavor. The advocates and practitioners of the "new philosophy" held that science should be presented in unadorned descriptive terms of experiment and observation, followed by strict inductions or deductions, in which each step was to be plain and clearly understood – without any rhetorical flourishes to distract the reader from the evidence and the logic. This was one of the reasons for the great esteem given to mathematics, which is perhaps the most rhetoric-free discourse imaginable.[105]

A classic example of metaphor – the assignment of a descriptive term to some object to which it is not strictly applicable – is the Scriptural comparison of life to a pilgrimage. Perhaps the most famous such metaphor is Shakespeare's comparison of life to a stage. Common metaphors include heart of flint, sharp mind, head of state, leaving no stone unturned, and eye of the law. A striking metaphor was used by James I soon after gaining the crown of England. "I am the husband," he told Parliament, "and the whole Isle is my lawful wife; I am the head and it is my body."[106]

The use of a metaphor does not necessarily imply any technical or scientific knowledge. When we use the metaphor of a marble brow, we mean only that we consider the flesh-and-blood brow to be cold and white like the brow of a marble statue; in this context we need not know anything about the chemistry or structural qualities of marble or the nature of the epidermis. But since a metaphor may also be based on erudition, a helpful distinction can be made between a popular or rough or untechnical metaphor and one that more learnedly invokes some element of the natural sciences. The difference between the two may be seen clearly by considering "body" in the metaphor of the body politic.[107] An example of a non-technical metaphor, one that does not involve the natural sciences, is found in 2 Corinthians, where St. Paul set forth a hierarchy of organs and parts of the body – from head and heart to limbs and belly – without any reference to medicine or physiology. It is the same for the oft-repeated Aesopian fable of the feet and the belly, in which the feet revolt because they believe they do all the work while the belly merely lies at ease above them doing nothing useful.[108] These examples may be contrasted with a statement in which James I likened the expanding metropolis of London to the spleen, "whose increase wastes the body." For here he was basing his metaphor on a physician's acquaintance with the function of the spleen. That is, he was invoking a resemblence between the operations of a city and the functions, considered technically, of an organ of the human body.[109]

All four levels of discourse may be discerned with respect to social applications of Newtonian physics.[110] First, there is the possibility of identity, a belief that the social world is a mechanical system operating under the same principles as the Newtonian system of the world.[111] Additionally, there were attempts, such as those we have seen made by Carey and Walras, to produce Newtonian homologues, laws in the social realm having the form of Newton's law of universal gravity; by contrast, Hume, Fourier, and Durkheim held only that they had produced a law which would have a function in a science of society that was an analogue of the function of the law of gravity in the Newtonian system. Others, however, merely believed that, on the level of metaphor, sociology or economics should be a "science" that, in some unspecified manner, would organize the subject in the way that Newton's *Principia* had done for the physical sciences. This was, apparently, the intent of Hamilton's *cri de coeur* of 1866:

> Although far more advanced, relatively, in particular ideas than sidereal philosophy before the time of Newton, it [social philosophy] scarcely less needs the PRINCIPIA MATHEMATICA PHILOSOPHIAE SOCIALIS, or rather the PRINCIPIA PRIMA.[112]

In explanation, Hamilton set forth what he believed to be the "Newtonian idea of Sociology," the assertion of

> the universality of the causes, or laws, which determine the social condition of mankind, and the consequent identity of the causes which determine the social destiny of an individual and a nation.

A variety of the Newtonian metaphor that has proved to be of significance for the social sciences consists of a Newtonian paradigm for social science based on Newton's method in general, using a procedure which I have called the Newtonian style.[113] This "style" does not refer to the set of mathematical techniques used by Newton – geometry and trigonometry, algebra, proportions, infinite series, and fluxions – but rather to the stages of contrapuntal interactions between imagined or ideal systems and those observed in physical nature.

The *Principia* begins with an idealized world, a mental construct comprising a single mathematical particle and a centrally directed force in a mathematical space. Under these idealized conditions, Newton can freely develop the mathematical consequences of the laws of motion which are the axioms of the *Principia*. At a later stage, after contrasting this ideal world with the world of physics, he will add further conditions to his intellectual construct: for example, by introducing a second body which will interact with the first one and then exploring additional mathematical consequences. Later, he will once again compare the mathematical realm to the physical world and revise the construct, for example, by introducing a third interacting body. In this way he can approach, by stages, nearer and nearer to the conditions of the world of experiment and observation, introducing bodies of different shapes and composition and finally considering bodies that move in various types of resistant mediums rather than in free space.

The *Principia* thus displays both the physics of an ideal world and the problems that arise because ideal conditions differ from the world of experience. For example, Newton shows that Kepler's first two laws of planetary motion are exactly true only for the mathematical or ideal condition of a single mass-point moving about a mathematical center of force, and he then develops the actual ways in which the pure form of Kepler's laws must be modified to fit the world of observation. The *Principia* can be accurately described as a work in which Newton explores, one by one, the ways in which ideal laws must be modified in the external world of experiment and observation.

A somewhat similar procedure was adopted in Thomas Malthus's *Essay on Population.*[114] Malthus stated a basic principle that "Population, when unchecked, increases in a geometric ratio." A later version says that "all animals, according to the known laws by which they are produced, must have the capacity of increasing in a geometrical progression."[115] This law is plainly not the result of a Baconian induction from a mass of observations. In fact, the law is true only of an unchecked population; a good part of Malthus's *Essay* is in fact devoted to evidence that populations do NOT so increase and to explanations of why they do not.

Malthus does not say that observed populations actually increase in a geometric or exponential ratio; he says explicitly that this *would* be the case for populations whose growth was not checked. The similarity of this statement and Newton's first axiom or law of motion will be immediately apparent. Newton did not write that all bodies move uniformly straight forward or stay at rest. Rather, he said that a body will maintain one or the other of those two "states" except to the extent to which impressed forces cause a change in state. Malthus is following the style of the *Principia* in seeking the reasons why the laws of the world of nature differ from those of the world of pure abstraction, in studying why real populations do not increase geometrically as they would in an ideal or imagined world.

In the *Essay* Malthus linked his presentation of the laws of population growth with Newton by citing Newton in terms of the highest respect even though Newton never wrote a word about populations or their increase. Malthus, we may note, excelled in mathematics and mathematical physics while an undergraduate at Cambridge, where he studied the *Principia* as well as commentaries on the Newtonian natural philosophy.[116] His use of Newton shows that the Newtonian natural philosophy has exerted a fruitful influence on the social sciences: not as a source of analogies or homologies, but in that metaphorical fashion which I have called the Newtonian style.

Those who have been engaging in critical-historical analysis of the social sciences, particularly economics, have not always made a clear distinction between analogy and homology, although the cases on which they have focused their critical attention – aspects of marginalist or neoclassical economics – exhibit examples of both. In particular, these analysts stress metaphor. In their usage, metaphor embraces both analogy and homology, but it may go even further to comprise the whole gamut of concepts, laws, theories, techniques, models, standards, and even values of the natural sciences (including mathematics) that economists have sought to borrow, emulate, imitate, or use in any way.

A close examination of the late nineteenth-century marginalist or neoclassical economists shows clearly the dual role of general metaphor and of specific analogy or homology. In the nineteenth century Isaac Newton still symbolized the highest level of scientific achievement, and the words used in relation to Newtonian science – "rational," "exact," and even "mathematical" – denoted a science at the zenith of the scientific hierarchy. Thus the emulation of Newtonian "rational mechanics" (complemented by the addition of such principles as those of Lagrange, d'Alembert, and Hamilton, together with such non-Newtonian concepts as energy) was an act of linking economics with the most successful branch of the natural sciences. This association was based on a metaphor. At the same time, however, rational mechanics provided concepts, principles, and even equations for which there seemed to be useful counterparts – both analogues and homologues – in economics.

Such explorations bring to our attention a very important aspect of interactions between the natural sciences and the social sciences, the transfer of value systems. William Stanley Jevons defended his attempts to introduce mathematics into economics by declaring that differential equations had been used traditionally in rational mechanics. In this statement Jevons was accomplishing two separate aims. He was justifying the introduction of the calculus into a social science and he was also implying that economics is like rational mechanics, then considered the paradigmatic exact science which all others should try to emulate. In short, he was implying that his subject shared the values of that branch of science which was then considered to represent the pinnacle of exactness and success.[117]

The value-laden aspect of metaphor often shows itself clearly in shifts of the specific part of the natural sciences which the social sciences are supposed to emulate. In this connection, the case of Jeremy Bentham is very illuminating. At different times in his life, he considered that the art-and-science of society should be modeled on medicine, even writing that "the art of legislation is but the art of healing practiced on a large scale," which – he added – was not a "mere fanciful" image. But at other times he chose as his paradigm the new chemistry, even conceiving that he would be its Lavoisier.[118] In one case he was lauding the beneficial practise of curing disease and maintaining health; in the other, the radical restructuring of knowledge.

We may see this feature of metaphor even more clearly in Engels's two eulogies of Karl Marx. At Marx's graveside, Engels's laudation took the form of a comparison with Darwin, indicating both the great effect of Darwin and Marx on contemporary thought and the revolutionary character of their ideas. Later, when he was editing Marx's *Nachlass* to produce the second volume of *Das Kapital*, Engels changed his metaphor for Marx's place in history. Now, as he wrote in his introduction, he found that Marx's counterpart was Lavoisier, the chief author (as Engels spent several paragraphs proving) of the Chemical Revolution.[119] While both Darwin and Lavoisier were symbols of scientific greatness, they represented quite different kinds of science and evoked metaphors comprising dissimilar sets of values and achievements. Both Darwin and Lavoisier were responsible for revolutions, but these were of very different types. Darwin radically altered our concept of species and their permanence, and his ideas challenged the existing order of thought in many fields of knowledge and belief. Lavoisier re-ordered the science of matter, and his work caused us to have a new and very different perspective on the constitution of substances, requiring that all substances, natural or synthetic, be given new names. Darwin turned an existing science upside down, but Lavoisier created a new science. Lavoisier made a legitimate science out of an old subject, just as, according to Engels, Marx had done in creating "scientific" economics.

Metaphors imply many aspects of the ways of doing science, the factors that must be considered whenever the historical or analytical focus is broad enough to encompass the total social and intellectual matrix in which science – whether natural or social – is done. Such considerations belong to the general historical interpretation of the sciences known commonly today as "external."[120] It has been argued that a primary reason why the "energy metaphor" was adopted by the neoclassical economists was not that it provided an accurate equivalent but that it invoked the values associated with the system of physics.[121] We are thus reminded that the choice of a particular metaphor to describe the interactions of the natural and the social sciences may suggest systems of values that are just as important as, or that may even be more important than, the compatibility of the concepts, principles, and quantitative elements.

1.6. ROLES OF ANALOGY

Analogies and similar types of correlation constitute a primary means of interaction between the natural sciences and the social sciences. These interactions are very much like those that occur between one branch of the natural sciences and another. They arise from a recognition that an idea, concept, law, theory, system of equations, method of investigation, mathematical tool, or any other element of one subject is similar to some element in another or has properties that enable it to be introduced usefully into that other subject. Analogy has always functioned as a tool of discovery, reducing a problem to another that has already been solved or introducing some element or elements that have proved their worth in a quite different area of knowledge. Jeremy Bentham once said that hints from analogies constitute one of the most important tools available for scientific discovery.[122]

A traditional use of analogy is to justify a novel or radical method or theory. An example would be the introduction of higher mathematics (e.g., the calculus) into economics on the analogy that the calculus had been used successfully in rational mechanics. A related use of analogy is to help explain abstruse concepts, as may be seen in all general presentations of relativity theory. Analogies also serve to make a difficult or strange idea seem reasonable and hence acceptable to the scientific community. An instance occurs in the work of Sigmund Freud.

Freud was hesitant about presenting in full one of his radical and difficult concepts, introduced only as a "suspicion" in 1890 in his *Interpretation of Dreams*. This was the idea that human beings have two different memory systems, one of which, as he wrote in 1924, "receives perceptions but retains no permanent trace of them," while the other preserves "permanent traces of the excitations" in " 'mnenomic systems' lying behind the perceptual system."[123] By 1924 he had discovered a mechanical device called the "Mystic Writing-Pad" (an older version of what in the United States is still called a magic slate) which seemed to simulate some main features of his concept. Emboldened by this encounter, Freud described his ideas about human memory in full, suggesting that the writing pad could be considered an analogue of his "hypothetical structure of our perceptual apparatus."[124]

Analogies were of significant importance in Freud's thinking and exposition. The "standard" edition of his collected works, in fact, contains a separate index for analogies. Best known of Freud's analogues are those which he drew from literature, notably Greek tragedy, in formulating and describing (and even naming) concepts. Freud was aware that in his cultural and anthropological studies – e.g., *Totem and Taboo* and *Moses and Monotheism* – "we are only dealing with analogies," and he fully recognized how dangerous it is, "not only with men but also with concepts, to tear them from the sphere in which they have originated and been evolved." It has been observed that by invoking an analogy Freud "likened religion to a collective obsessional neurosis, or allowed that Hamlet suffered unduly from an Oedipus complex."[125]

The explicit use of analogies was introduced into science during the formative years of the Scientific Revolution. In an extensive study of this subject, Brian Vickers has found that in the late Renaissance and early seventeenth century, the attitude toward analogies constituted a major issue on which the new science diverged from an occult tradition.[126] The new science, according to Vickers, stressed a "distinction between words and things and between literal and metaphorical language." In the occult tradition, however, words were "treated as if they are equivalent to things and can be substituted for them." As a result, analogies were not, as they were "in the scientific tradition, explanatory devices subordinate to argument and proof, or heuristic tools to make models that can be tested, corrected, and abandoned if necessary"; they were, instead, "modes of conceiving relationships in the universe that reify, rigidify, and ultimately come to dominate thought." I would modify this conclusion only to the degree of adding that for scientists analogy also served as an instrument of discovery.

One of the early scientists to make extensive use of analogies was Johannes Kepler, who wrote, in his epoch-making work on optics, "I especially love analogies, my most faithful master."[127] In the same work, Kepler indicated how analogy is used in the process of discovery: "Analogy has shown, and Geometry confirms." He employed analogy especially in his *Astronomia Nova* in 1609, where he set forth his first two laws of planetary motion. Reasoning, as he said, "by analogy," Kepler made use of the properties of such "intangibles" as light and magnetic force in order to develop the idea of a solar (or "solipetal") force acting on the planets. He was clear about the distinction between analogy and identity, even stating, with respect to his postulated magnetism of planets: "Every planetary body must be regarded as being magnetic, or *quasi*-magnetic; in fact, I suggest a similarity and not an identity."[128]

Newton also reasoned in terms of analogy and even formalized the use of analogies in natural science in his *Principia*, in the second of what he called the "Regulae Philosophandi" or "Rules of Natural Philosophy." The "causes to be assigned to natural effects of the same kind should" he wrote, "be so far as possible, the same." The examples he gave were "respiration in man and beast," "the falling of stones in Europe and America," "the light of a kitchen fire and of the sun," and the "reflection of light on our earth and in the planets."[129]

A comparable way in which analogies serve science is in exhibiting the validity of a conclusion that seems untestable. In discussing the stability of the solar system in his *Système du monde*, Laplace had to argue that certain observed variations are not secular but periodic; they seem to be secular only because they have a period extending over millions of years. Laplace showed that the system of Jupiter's satellites in a dynamical analogue of the solar system, the satellites displaying in their motions the same perturbations as the planets. Since the satellites exhibit all the phases of their mutual gravitational perturbations within a few centuries, the periodic nature of the oscillations can be verified, thus making it likely by analogy that the similar variations in the planetary motions are also periodic.[130]

Both Charles Darwin and his contemporary James Clerk Maxwell made frequent use of analogies. Darwin's basic concept of a "struggle for existence" was presented in the *Origin of Species* (1859) on the basis of analogy with Malthus's principles of population. Malthus's two laws dealt only with human populations, which if unchecked would naturally increase in an exponential ratio. By analogy Darwin inferred that all populations of organic beings – human, animal, plant – would naturally increase exponentially so that, as he wrote, "more individuals are produced than can possibly survive," with the result that "there must be a struggle for existence." The analogy, Darwin noted, was not exact since, in the natural plant and animal world – unlike the human world of agriculture – "there can be no artificial increase in food." Nor in the plant and animal world is there exercised that "prudential restraint from marriage" which Malthus found to exert a moral rein on human population growth.[131]

Maxwell not only made extensive use of analogies but wrote eloquently about their role in science. His discussions of analogy remain today perhaps the best introduction to this subject.[132] One example of his use of analogies occurs in relation to the theory of heat. The "laws of the conduction of heat in uniform media," he wrote, "appear at first sight among the most different in their physical relations from those relating to attractions." Even so, he concluded, we "have only to substitute *source of heat* for *centre of attraction*, *flow of heat* for *accelerating effect of attractions* at any point, and *temperature* for *potential*," and the result is that "the solution of a problem in attractions is transformed into that of a problem of heat." So exact is the formal analogy that "if we knew nothing more than is expressed in the mathematical formulae, there would be nothing to distinguish between the one set of phenomena and the other" – despite the fact that the conduction of heat "is supposed to proceed by an action between contiguous parts of a medium, while the force of attraction is a relation between distant bodies."[133] Having justified this method, Maxwell proceeded to use it in elaborating a mathematical theory of "lines of force" (in Faraday's sense) by making use of an analogy with the "mathematical formalism" of the motion of an incompressible and imponderable fluid.[134]

1.7. RATIONAL MECHANICS AND MARGINALIST ECONOMICS

In considering the role of analogies and similar correlations in the social sciences, two primary areas of nineteenth-century natural science attract our attention. Mathematical physics, consisting of the new rational mechanics plus energy physics, had a profound influence on economics, while the cell theory, together with related aspects of the life sciences, gave new form as well as content to theories of social morphology and behavior.

These two subject-areas illustrate very different aspects of the ways in which social sciences draw on the natural sciences. Rational mechanics with energy physics provided a rich source of conceptual homologues for a rising marginalist (or neoclassical) economics, together with analytical tools such as Lagrangian virtual displacements and Hamiltonian functions, even analogous equations and principles of minimization and maximization. While producing a social science with an external appearance of physics, some of the founders of neoclassical economics wholeheartedly adopted the metaphor of mathematical physics, clearly hoping to give the social science of economics a legitimation (especially in the opinion of natural scientists) and some measure of the value-system of "hard" science.[135] Economists of this school have continued to draw on the body of physics well established by the end of the nineteenth century. Apparently, they have felt little need to encompass within their theoretical structures any later developments such as quantum theory or relativity. An outsider cannot help but be astonished that economics has been affected so little by the later dramatic revolutions in the very subjects – rational mechanics and energy physics – which have provided some of its principal metaphors. For example, there seems to be no current significant economic ripple from the twentieth-century conclusions that the conservation of energy can no longer be considered an independently true principle and that energy itself can no longer be regarded as subject to continuous variation but acts in quantized steps. Perhaps this paradox is to be explained by the judgment of Philip Mirowski and other critics of neoclassical economics that the energy metaphor was only imperfectly understood by the founders, who apparently were not aware that their adopted energy model was flawed because they did not take account of the conservation law.

Ernest Nagel has divided analogies into two classes: "formal" and "substantive." A "substantive" analogy is one in which a theory or a system is patterned on the model of another system which contains known laws.[136] Examples are the kinetic theory of gases (patterned on the known laws of the interaction of elastic spheres such as billiard balls), electron theory (in which the analogy is with macroscopic electrostatically charged bodies), and atomic structure (the model is the solar system). The other type of analogy is "formal," based on a structure of abstract relationships rather than a "more or less visualizable set of elements." An example would be the analogy proposed by J. C. Maxwell based on the isomorphism of the laws of gravitation theory and the laws of heat conduction.[137] Neoclassical economics illustrates the use of such formal analogies.

This kind of example from economics, however, goes far beyond a mere creative transfer of concepts and principles, mathematical expressions, and other tools of the arsenal of mathematical physics. Economics, and to some degree the other social sciences, may illustrate a thesis of Jevons that analogy "leads us to discover regions of one science yet undeveloped, to which the key is furnished by the corresponding truths in the other science."[138] To make this sentiment universally valid, we should enlarge Jevons's "corresponding truths" by adding methods and formal techniques (e.g., equations).

Economics and mathematical physics seem at first sight to be extremely different. Economics deals with such human and moral or ethical factors as greed, profit, cost, value, utility, need, and good. These topics appear to be worlds apart from such abstractions as force, field, distance, speed, and kinetic and potential energy; they appear to be free of affect and seem to lend themselves "naturally" to mathematical treatment. But analogies between very different subjects are not unusual in the history of science: "No two sciences might seem at first sight more different in their subject matter than geometry and algebra," Jevons wrote, since one deals with "forms in space" (circles, squares, triangles, parallelograms, . . .) and the other with abstract "symbols and numbers."[139] Yet, as Jevons pointed out, a crucial step in the development of modern mathematics was the recognition of analogies between these two branches of mathematics. He described Descartes's great breakthrough as a demonstration of a "most general kind," that equations may be represented by curves or figures in space and vice versa and "that every bend, point, cusp or other peculiarity in the curve indicates some peculiarities in the equation." Jevons found it "impossible to describe in any adequate manner the importance of this discovery."[140]

This kind of analogy occurs frequently in the social sciences, notably in economics. In his *Theory of Political Economy*, Jevons took note of the "objections made to the general character of the [differential] equations" which he had employed, defending his position by making an analogy between economics and physics, declaring that economics is similar to physics insofar as "the equations employed do not differ in general character from those which are really treated in many branches of physical science."[141] The example he chose to develop was the use of the principle of virtual velocities (or virtual displacements) applied to the lever, where there is a homology of equations, that is, the equations for the case of the lever "have exactly the forms of the equations [in economics]." He even introduced a diagram in order to "put this analogy of the theories of exchange and of the lever in the clearest possible light."[142] This same kind of analogy of theories was invoked by Léon Walras in an article on analogy, "Economique et mécanique," published in 1909. Here Walras argued that identical differential equations appear in his analysis of economics and in two examples from mathematical physics: the equilibrium of a lever and the motion of planets according to gravitational celestial mechanics.[143] Claude Ménard has described Walras's text on "Economics and Mechanics" as a term-by-term comparison of the proportion between *rareté* (scarcity, i.e., marginal utility) and value – which is the basis of the theorem of maximum satisfaction – with the equation of maximal energy from rational mechanics. In addition, Ménard indicates that Walras' law, defining the properties of general equilibrium in relation to the marketing of goods, services, and money, relies on the example of uniformly accelerated motion from celestial mechanics and invokes equations containing mass and acceleration.[144]

Vilfredo Pareto was writing as an economist when he invoked a similar "formal" homology in the example of "the equations which determine [economic] equilibrium." On seeing these equations, he wrote, a writer trained in mathematical physics (as he was) would observe, "These equations do not seem new to me; I know them well, they are old friends. They are the equations of rational mechanics." Because the equations are the same, he concluded, "pure economics is a sort of mechanics or akin to mechanics."[145]

Pareto envisioned a double role for mathematics in economics and more generally in social science. Mathematics, he believed, provides a means of analogically transferring the basic equations of physics to economics. Mathematics also serves as a primary tool for dealing with such problems as the "mutual dependence of social phenomena" in conditions of equilibrium; here mathematical analysis enables us to make precise "how the variations of any one of these [conditions] influence the others," an assignment in which "we really need to have *all* the conditions of the equilibrium." In the "existing state of our knowledge," he noted, only mathematical analysis can "tell us if this requirement is observed."[146]

This led Pareto to some remarks on the proper role of analogies and the dangers of using them in social science. Since "the human intellect proceeds from the known to the unknown," he wrote, we can make progress in our thinking by basing our ideas of an area of the "unknown" on analogies drawn from an area of the "known." For example, "extensive knowledge of the equilibrium of a material system," helps us to "gain a conception of economic equilibrium" and this in turn "can help us to form an idea of social equilibrium." He warned, however, that in "such reasoning by analogy there is . . . a pitfall to be avoided." That is, the use of analogies "is legitimate, and perhaps highly useful, as long as what is involved is only the elucidation of the sense of a given proposition." We are led into grave errors, however, if we try to use analogies to prove a proposition or even "establish a presumption in its favour." Analogies, he added, serve primarily to clarify the meaning of propositions.[147]

Philip Mirowski has devoted a good part of his book, *More Heat Than Light*, to an argument that such figures as Jevons, Walras, Edgeworth, Fisher, and Pareto – all leading architects of the Marginalist Revolution – based their economics on, or at least associated it with, the mathematics of a specific subset of physics: post-Newtonian rational mechanics (i.e., incorporating principles of Lagrange and Laplace plus the methods of Hamilton) combined with the doctrines of energy. There was thus conceived, on the level of metaphor, a correspondence between economics and physics. And even before the marginalist school of economics had come into being there were expressions of hope that economics might become a true or exact science on the model of mathematical physics. In 1875 this position was expressed clearly by J. E. Cairnes: "Political Economy is as well entitled to be considered a 'positive science' as any of those physical sciences to which this name is commonly applied." The principles of economics, he asserted, are "identical" in character "with that of the physical principles which are deduced from the laws of gravitation and motion."[148] For Jevons, the emerging new economics was regarded as making use of concepts which were direct homologues of physical concepts. With a sense of security coming from the use of equations homologous to those in physics, the new economics assumed the metaphor of rational mechanics and its great founder Isaac Newton, including scientific dignity, precision, esteem, and the whole value system. It is difficult for us today to imagine or reconstruct the veneration in which nineteenth-century scientists held Isaac Newton and his law of universal gravity, but we may gain a hint of the awe inspired by Newton and his law by considering a moment in the life of Charles Kingsley. In 1860, when his son had just died and when it seemed to him that all his foundations of faith were crumbling, the bereaved parent wrote to Thomas H. Huxley: "I know what I mean when I say I believe in the law of the inverse square, and I will not rest my life and my hopes upon weaker convictions."[149] The Newtonian law of gravity, together with the laws of motion, provided a certainty on which intellectuals could agree.

Some founders of the Marginalist Revolution believed in a homology of concepts in economics and in physics, with the consequence that the laws of one could be directly translated into the other. Jevons, for example, stated *expressis verbis* that the "notion of value is to our science what that of energy is to mechanics." He even adopted directly from Maxwell the technique of dimensional analysis (*L*, *T*, and *M*: length, time and mass) and showed that "the *dimensions of commodity*, regarded merely as a physical quantity, will be *the dimensions of mass*." The homology extended ultimately to Newton's law of gravity when Jevons declared that "utility is an attraction between a wanting being and what is wanted" and is "just" like "the gravitating force of a material body."[150]

In a similar vein, Léon Walras later wrote in his *Elements of Pure Economics* that the use of "mathematics promises to convert pure economics into an exact science," that "mathematical economics will rank with the mathematical sciences of astronomy and mechanics." He concluded that "the pure science of economics is a science which resembles the physico-mathematical sciences in every respect."[151] To put this outlook into perspective, Claude Ménard has argued that Walras "sought justification and guarantee as much as inspiration" in his use of the analogy between economics and rational mechanics. In this regard, Ménard stresses the fact that Walras was "always concerned with scientific legitimacy and despairing of recognition of the value of his work."[152] Moreover, in a scholarly study of Walras's economic ideas in relation to Mirowski's analysis, Albert Jolink denies "strenuously" that Walras's economics theory "slavishly imitates physics." Jolink cites evidence that Walras had little or no understanding of energy physics before 1906, and he concludes that even after 1906 it is doubtful whether

TABLE I. Pareto's Analogies

Mechanical Phenomena	Social Phenomena
Given a certain number of material bodies, the relationships of equilibrium and movement between them are studied, any other properties being excluded from consideration. This gives us a study termed *mechanics*. This science of mechanics is divisible into two others:	Given a society, the relationships created amongst human beings by the production and exchange of wealth are studied, any other properties being excluded from consideration. This gives us a study termed *political economy*. This science of political economy is divisible into two others:
1. The study of material points and inextensible connections leads to the formulation of a pure science – rational pure mechanics, which makes an abstract study of the equilibrium of forces and motion. Its easiest part is the science of equilibrium. D'Alembert's principle enables dynamics to be reduced to a problem of statics.	1. The study of *homo economicus*, of man considered solely in the context of economic forces, leads to the formulation of pure political economy, which makes an abstract study of the manifestations of ophelimity. The only part we are beginning to understand clearly is that dealing with equilibrium. A principle similar to D'Alembert's is applicable to economic systems; but the state of our knowledge on this subject is still very imperfect. Nevertheless, the theory of economic crises provides an example of the study of economic dynamics.
2. Pure mechanics is followed by applied mechanics which approaches a little more closely to reality in its consideration of elastic bodies, extensible connections, friction, etc. Real bodies have properties other than mechanical. Physics studies the properties of light, electricity and heat. Chemistry studies other properties. Thermodynamics, thermochemistry and the like sciences are concerned specifically with certain categories of properties. These sciences all constitute the physico-chemical sciences.	2. Pure political economy is followed by applied political economy which is not concerned exclusively with *homo economicus*, but also considers other human states which approach closer to real man. Men have further characteristics which are the object of study for special sciences, such as the sciences of law, religion, ethics, intellectual development, esthetics, social organisation, and so on. Some of these sciences are in an appreciably advanced state; others are extremely backward. Taken together they constitute the social sciences.

TABLE I. *(Continued)*

Mechanical Phenomena	Social Phenomena
Real bodies with only pure mechanical properties do not exist.	Real men governed only by motives of pure economics do not exist.
Exactly the same error is committed *either* by supposing that in concrete phenomena there exist solely mechanical forces (excluding, for example, chemical forces), *or* by imagining, on the other hand, that a concrete phenomenon can be immune from the laws of pure mechanics.	Exactly the same error is committed *either* by supposing that in concrete phenomena there exist solely economic motives (excluding, for example, moral forces), *or* by imagining, on the other hand, that a concrete phenomenon can be immune from the laws of pure political economy.
The difference between practice and theory arises precisely from the fact that practice has to take account of a mass of details which theory does not deal with. The relative importance of primary and secondary phenomena will differ according to whether the viewpoint is that of science or of a practical operation. From time to time, attempts are made to synthesise all the phenomena. For example, it is held that all phenomena can be ascribed to:	
The attraction of atoms. The attempt has been made to reduce to unity all physical and chemical forces.	Utility, of which ophelimity is only a type. The attempt has been made to find the explanation of all phenomena in *evolution.*

"Walras had any understanding at all concerning a proto-energetic metaphor."[153] In short, for Walras the physical analogues served more as a means of later legitimation of his economics than as a primary instrument of discovery. Yet there can be no doubt that Walras wished to associate his economics with mathematical physics.

Pareto was equally convinced that the "equilibrium of an economic system offers striking similarities with that of a mechanical system," but he was aware that there are special pitfalls for those who study political economy without "a knowledge of pure mechanics." Firm in his conviction that an analysis of a mechanical system is of the greatest help in giving "a clear idea of the equilibrium of an economic system," he drew up a table (printed here as Table I) for "those who have not studied pure mechanics" and who will need help in understanding the argument. In this table he placed in parallel columns some major concepts and principles of physical mechanics and their counterparts in economics. He warned, however, that in such a tabulation of "analogies existing between mechanical and social phenomena" the "analogies do not prove anything: they simply serve to elucidate certain concepts which must then be submitted to the criterion of experience."[154]

The extreme of this proposed homology between economics and rational mechanics is found in Irving Fisher's *Mathematical Investigations into the Theory of Value and Prices* (1926). It should be noted that Fisher was rather well trained in mathematics and physics (as Jevons and Walras were not), having been one of the small group of students who worked for their Ph.D. under J. Willard Gibbs. In the style of Pareto, with whom he was in correspondence, Fisher (see Table II) also drew up a table of homologies from physical mechanics and economics. His compilation, however, goes beyond Pareto's to the extent of including not only paired concepts (such as particle and individual; space and commodity; energy and utility) but also the property of being scalar or vector, and his list was extended to include even general principles.

Philip Mirowski found, however, that despite Fisher's parade of dynamical analogies and homologies, he apparently took "most of his analogies . . . from hydrostatics rather than from fields of force." Mirowski notes, in this regard, that in an unpublished essay on "My Economic Endeavors" Fisher boasted of having pioneered in "hydrostatic and other mechanical analogies." Mirowski has presented a critique of Fisher's table, beginning with the "incorrect" identification "of a particle with an individual." Like other "neoclassical economists," Fisher – according to Mirowski's thesis – made a serious blunder in not appreciating the principle of conservation of energy, which would imply for an economic system "that the sum of total expenditure and the sum of total utility in a closed trading system must be equal to a constant." Mirowski argues that Fisher's general failure to carry the physical analogy to its logical conclusion – that is, to take cognizance of the conservation law – was a logical fault that came from an incomplete understanding of the physics metaphor of energy and field that lies at the very foundation of neoclassical economics.[155] It must be admitted, however, that all economists do not accept this radical critique.[156]

One of the difficulties in using analogies, whether in the natural sciences or in the social sciences, is that there may be more than one analogy for the same problem. The problem of multiple analogies, along with the concomitant need for a decision concerning which one to choose, has long plagued the social sciences. It arose in a dramatic fashion in 1898 in Alfred Marshall's discussion of "Mechanical and Biological

TABLE II. Fisher's Analogies

Mechanics	Economics
a particle	an individual
space	commodity
force	marginal utility or disutility
work	disutility
energy	utility
work or energy = force × space	utility = marginal utility × commodity
force is a vector	marginal utility is a vector
forces are added by vector addition	marginal utilities are added by vector addition
work and energy are scalars	disutility and utility are scalars
The total energy may be defined as the integral with respect to impelling forces.	The total utility enjoyed by the individual is the like integral with respect to marginal utilities.
Equilibrium will be where net energy (energy minus work) is maximum; or equilibrium will be where impelling and resisting forces along each axis will be equal.	Equilibrium will be where gain (utility minus disutility) is maximum; or equilibrium will be where marginal utility and marginal disutility along each axis will be equal.
If total energy is subtracted from total work instead of vice versa the difference is "potential" and is a minimum.	If total utility is subtracted from total disutility instead of vice versa the difference may be called "loss" and is minimum.

Analogies in Economics."[157] After a discussion of dynamics and statics in relation to economics, Marshall – who was well trained in physics and mathematics – expressed deep skepticism about the analogy with physics. He concluded that while "there is a fairly close analogy between the earlier stages of economic reasoning and the devices of physical statics," there is not "an equally serviceable analogy between the later stages of economic reasoning and the methods of physical dynamics." At the later stages, he argued, "better analogies are to be got from biology than from physics." Accordingly, "economic reasoning should start on methods analogous to those of physical statics, and should gradually become more biological in tone." This need for shifting analogies was apparently very important for Marshall. Analogies, he wrote, "may help one into the saddle, but are encumbrances on a long journey." That is, it is "well to know when to introduce them, it is even better to know when to stop them off." He concluded that "in the later stages of economics, when we are approaching nearly to the conditions of life, biological analogies are to be preferred to mechanical."[158] On the title page of Marshall's *Principles of Economics* there is a biological apothegm taken directly from Darwin's *Origin of Species*: "Natura non facit saltum."

1.8. BIOLOGICAL THEORY AND SOCIAL THEORY

The situation with respect to the "organismic" theories of society is quite different from that of marginalist or neoclassical economics. The sociologists, unlike the economists, gloried in revealing the sources of their analogies and homologies and other comparisons and correlations and in showing how current their biological knowledge was. They even went to the extent of inserting in their sociology biological tutorials on the latest development. In three cases examined below, those of Lilienfeld, Schäffle, and Worms,[159] we can see the joy and satisfaction derived from using the latest findings in biology. This trio of thinkers shared the historical recognition that the cell theory had brought the life sciences to a state of maturity – a conclusion which led to hopes that the use of the cell theory would produce a similar effect in sociology. One can trace in their sociological works the successive ideas of von Baer on embryological development and the increase of complexity as a part of development, the doctrine of Milne-Edwards and others on division of labor in relation to cell function and structure, Virchow's cellular pathology, and the new ideas relating to the germ theory of disease.

Organismic sociology and marginalist economics differ even further in a number of fundamental respects. Some founders of the Marginalist Revolution (e.g., Jevons and Walras) were deficient in their actual understanding of the mathematical physics which they claimed their subject to be emulating, while the proponents of the organismic theories of society had a sound grasp of biological principles – perhaps an easier assignment than to understand physical principles. The greatest differences between the two groups, however, is that neoclassical economics still flourishes as a dominant school of thought, whereas the organismic theories of society have largely withered away and may even seem ridiculous to today's reader. As a result, the preceding material on economics appears to be a study of the founding period of today's thought, whereas the ideas of the organismic sociologists seem so extravagant that many historical examinations of the writers of this school end up as total disparagements.[160]

Why was there in the late nineteenth century so vigorous a school of social thought based on an exact parallel with the life sciences? In order to understand why, we must take into account the great achievements of the biological sciences, together with the extraordinary successes of medicine, in the nineteenth century. This century witnessed tremendous advances such as the cell theory and the theory of evolution plus developments in embryology, physiology, and morphology that completely transformed the subject. The new science of microbiology had not only opened up an exciting new realm of biology but at last provided medicine with a knowledge of the causes of contagious diseases and even showed the way to prevent or to cure some of them. Great new prospects seemed in store for the life sciences: conquest of yet additional diseases, finding the key to the origins of life, understanding the processes of heredity, and much more. By contrast, physicists were sounding the gloomier message of making more exact measurements of the constants of nature, even expressing a conviction that the future was to be found in the next decimal place. We may easily understand why many social scientists of the late nineteenth century could believe that a new great age of biology was taking the place of the older great age of physics. This point of view was expressed dramatically by the economist Alfred Marshall in his "Inaugural Lecture" at Cambridge University in 1885. "At the beginning of the nineteenth century," he said, "the mathematico-physical group of sciences was in the ascendant." But now "the speculations of biology [have] made a great stride forwards." The discoveries in biology, he continued, now attract "the attention of all men as those of physics had done in earlier years." The result was that the "moral and historical sciences of the day have . . . changed their tone, and Economics has shared in the general movement."[161]

Additionally, not only were social scientists impressed by the achievements of biological science, but many sociologists were convinced, as Comte had taught explicitly, that because sociology deals with human behavior it must be a science very close to, or very much like, biology. There is, accordingly, no reason to wonder why the organismic sociologists chose to construct a science in emulation of biology, and as historians we may focus our attention on the degree to which they were successful in finding relevant analogues and homologues for producing a biological science of society. The important bio-medical subjects used by sociologists included the cell theory; the new embryology; the physiology centering on the *milieu intérieure*; cellular pathology; the germ theory of disease; new theories concerning psychological disorders, notably hysteria;[162] and, of course, the theories of evolution.[163]

Most organicists traced a lineage that went back to Auguste Comte. Although he is usually thought of primarily in the context of social physics, Comte also made extensive use of the organismic metaphor, drawing heavily on physiology and pathology. In his *Course of Positive Philosophy* he clearly set forth the important point of view that social disturbances should be regarded as pathological cases, being, "in the social body, exactly analogous to diseases in the individual organism." Comte held the extreme position that in the development of biological science, "pathological cases are the true equivalent of pure experimentation." It followed that the study of social pathology should provide the equivalent of social experiment, something which he was aware can never occur to the same degree and kind as in physics or chemistry.

Comte esteemed and drew heavily on the ideas of Broussais, one of the great reformers of medicine; in his *System of Positive Polity* (1848–1854), Comte wrote of " . . . the admirable axiom of Broussais" which "destroys the old absolute distinction between health and disease." Between these extreme limits, Comte added, "we may always find a multitude of intermediate stages, not merely imaginary, but perfectly real, and together forming an almost insensible chain of delicate gradations."[164] Broussais taught Comte that pathology, "the study of malady, is the way to understand the healthy state." Primarily it was his "principle of continuity" which guided Comte's own analysis: "that the phenomena of the pathological state are a simple prolongation of the phenomena of the normal state, beyond the ordinary limits of variation." Until now, Comte declared, no one had drawn the analogy between physiological and social pathology, no one had ever applied "this principle to intellectual and moral [i.e., social] phenomena."[165]

At the century's end, in 1896, the American cytologist Edmund Beecher Wilson boldly declared the cell theory to be the second great generalization made by biology, the first having been organic evolution.[166] In retrospect, insofar as social theory or social science is concerned, the cell theory seems to have been at least of equal importance with the Darwinian evolutionary theory. It is easy to see why the cell theory and its sequelae had so great an impact on social science. As many were quick to observe, the concept of a natural organism as an organized system of living cells provided a new scientific foundation for an organismic concept of society. The cells resemble the individual members of human society in the degree to which each cell has a life of its own while the lives of all the cells are linked together. Additionally, the cells in the animal or human body exhibit the principle of the physiological division of labor, since each type of cell has a structure especially adapted for its function within the organism: nerve cells to transmit messages or commands, liver cells to produce bile, and so on.[167] This principle became central to the biological thought of Milne-Edwards and others and from them passed at several removes to Emile Durkheim, who applied it within a sociological framework in his major doctoral dissertation. Furthermore, cells are grouped together into functional units (tissues, organs) just as human individuals are organized into social units. Even the distribution or circulation of nutriments and the discharge of waste products could be seen analogically in natural bodies composed of cells and in social bodies composed of humans.

The analogy ultimately broke down because, although each cell has a life of its own, no body-cell can survive on its own apart from the parent body-matrix. Remove a muscle cell and it will, of and by itself, quickly die. Additionally, there is the problem of "will," which exists in each human being in society and which has no counterpart in individual cells.[168] Furthermore, the animal body differs from the social organism in having a relatively short and determinate lift-span and in exhibiting a regular sequence of universally accepted and recognized symptoms of decay and old age.[169] Nevertheless, the perceived similarities continued to be important in social science.

The significance of the cell theory for a science of society was enhanced by the embryological discoveries of Karl Ernst von Baer and his successors. The recognition of the stages of development of the embryo by cell division from a single cell, and the subsequent elaboration of organ and tissues, suggested a similar sequence of social organization, starting from a single mother (as the original cell) and, by subsequent multiplication, accompanied by grouping of individuals (similar to the grouping of cells), forming family units, then tribes, and eventually countries.[170]

Of special importance for social scientists was von Baer's principle that the stages of development form a sequence characterized by a transition from simplicity to greater and greater complexity. This was similar to the discovery that extinct and living forms of animals could be ordered in an ascending scale of development in which there was an increasing degree of complexity.[171] In its most complete form, this result became encapsulated in the famous "biogenetic law," that "ontogeny recapitulates phylogeny," which preceded Darwin's theory of evolution and was found to be conformable to either Darwinian or Lamarckian evolution.[172]

Social scientists such as Herbert Spencer[173] drew upon von Baer's results to formulate an analogous developmental theory for society, conceived to have gone through a regular sequence from infancy to old age just like the development of mankind from savagery to civilization.[174] There would thus be a general law of evolution that applies to the development of animals from earliest times, to the development of the embryo, to the development of civilization, and to the development of societies. In a piece first published in 1864,[175] Spencer explained that his thought had been profoundly influenced by "the truth that all organic development is a change from a state of homogeneity to a state of heterogeneity" and that it was von Baer who had given this truth "a definite shape." Incorporating von Baer's "formula" into his own "beliefs in evolutions of various orders," Spencer had further extended and modified both von Baer's concept and his own insights, thus involving his thought in a process "of continuous development, set up by the addition of von Baer's law to a number of ideas that were in harmony with it."[176]

An aspect of the cell theory which had special importance for nineteenth-century organismic sociology, and especially for consideration of the social analogues of the cell theory, was introduced by Rudolf Virchow's doctrine of "cellular pathology" (1858).[177] This doctrine held that all pathological conditions of the human body could be attributed to a state of degeneration or a condition of abnormal activity of some individual constituent cell or cells. Thus Virchow transformed thinking about the body as a whole to thinking about conditions of the fundamental biological units of which the body is composed. One important consequence of Virchow's doctrine was that pathological conditions were seen as merely extremes of the normal rather than as different in kind. In the present context his ideas are especially interesting also because the biologist himself stressed the similarities between biological phenomena and sociological phenomena. According to Virchow, just as a tree constitutes a mass arranged in a definite manner, in which, in every single part, in the leaves as in the root, in the trunk as in the blossom, cells are discovered to be the ultimate elements, so is it also with the forms of animal life. *Every animal presents itself as a sum of vital unities*, every one of which manifests all the characteristics of life. The characteristics and unity of life cannot be limited to any one particular spot in a highly developed organism (for example, to the brain of man), but are to be found only in the definite, constantly recurring structure, which every individual element displays. Hence it follows that the structural composition of a body of considerable size, a so-called individual, always represents a kind of social arrangement of parts, an arrangement of a social kind, in which a number of individual existences are mutually dependent, but in such a way, that every element has its own special action, and, even though it derives its stimulus to activity from other parts, yet alone effects the actual performance of its duties.[178]

Believing that all plants and animals are aggregates of cells as the fundamental life-units, Virchow concluded that all structural and functional properties of organisms are determined by relations among individual cells.[179] In referring to the cells as providing the "living organism" with a "multiplicity of vital foci," Virchow explained that every organism

is a free state of individuals with equal rights though not with equal endowments, which keeps together because the individuals are dependent upon one another and because there are certain centers of organization with whose integrity the single parts cannot receive their necessary supply of healthful nourishing material.[180]

As Owsei Temkin has indicated, "the metaphor of the cell state for Virchow was not a mere manner of speech, but an integral part of his biological theory."[181] Here we note a striking example of the use of social concepts in the thought of a biologist.

Virchow provided a direct model for such social scientists as Lilienfeld and Schäffle. Of particular importance in this respect is Paul von Lilienfeld's[182] *Social Pathology* (1896), which must be read in the light of his five-volume opus, *Thoughts on the Social Science of the Future* (1873–1881). In the earlier work, at the beginning of the first volume, *Human Society as a Real Organism*, Lilienfeld issued his challenge:

Human society is, like natural organisms, a real being, is nothing more than a continuation of nature, is only a higher expression of the same forces that underlie all natural phenomena: This is the assignment, this is the thesis, which the author has set himself to accomplish and to prove.[183]

In *Social Pathology* he continued the challenge by asking how the study of society could be made truly scientific and by giving the solution which he felt he had demonstrated in the *Thoughts*:

The condition *sine qua non* by which sociology may be raised to the rank of a positive science and by which the inductive method may be applied to it is . . . the conception of human society in its character as a real living organism, composed of cells as are the individual organisms of nature.[184]

He went on to identify the cellular structure of society: "Social cells are human individuals forming first the family, then the clan, the tribe, the nation," and finally, at present, the state, and probably, in the future, humanity as "a great organic whole."[185]

In developing the "physiology and morphology" of the human social organism, Lilienfeld gave primacy to a "constituent factor of every human association, beginning with the family and going up to the state and the whole human race." This is the "nervous system, the source of all social action."[186] He argued that "the intercellular substance of individual organisms" corresponds in the social organism to "the wealth produced, exchanged, and consumed." He provided evidence to support his conclusion that any action of the social nervous system on this ambient milieu is very much like the physiological action of individual organisms endowed with nervous systems. He further insisted that in the social organism there are specific nerve energies just as in natural organisms, "not only in a figurative sense, but really."[187]

Lilienfeld hailed Virchow's "cellular pathology" as "one of the most striking conquests of modern science." From Virchow he had learned that "every pathological state of the human body derives from a degeneration or an abnormal activity of the simple cell, as the elementary anatomical unit from which every organism is constructed."[188] Furthermore, Lilienfeld wrote, Virchow had taught that "there is no essential and absolute difference between the normal state and the pathological state of an organism." In a "deviation from the normal state," Lilienfeld declared (on Virchow's authority), "a cell or a group of cells manifests an activity outside the necessary time, outside the necessary place, or outside the limits of excitation prescribed by the normal state."[189]

As every individual disease derives from a pathological state of the cell, likewise every social disease has its cause in a degeneration or abnormal action of the individual who constitutes the elementary anatomical unit of the social organism. Likewise, a society attacked by disease does not present a state essentially different from that of a normal society. The pathological state consists only in the manifestation by an individual or a group of individuals of an activity that is untimely or is out of place or indicates over-excitement or lack of energy.[190]

While there is no doubt that Virchow's "cellular pathology" revolutionized medicine, it had a fundamental weakness in that it did not take account of communicable diseases, a topic illuminated by the germ theory of disease. Lilienfeld was aware of this deficiency and accordingly supplemented Virchow's analysis by introducing the discoveries concerning the germ of disease. "It has now been proved," he wrote, "that to each disease there corresponds a specific bacillus." As we would expect, Lilienfeld identified certain social diseases as similarly caused by "specific parasites." The "social organism is infested with economic, juridical, and political parasites," subdivided "into several classes and species, each one of which corresponds to a special disease of the social organism."[191]

Lilienfeld argued that there are even more fundamental and remarkable connections between living organisms and society. He held that "organic nature itself presents three degrees of development and perfection." The first is that plants cannot move autonomously, either together as a whole or separately as parts. The second: animals can move freely, but only as individuals, that is, as parts. But, third, a "social aggregate" can move freely both as a whole and in its parts. Thus "it is only in human society that nature realizes in its fullness the highest degree of organic life: the autonomy of the same individual organism in the parts and in the whole."[192]

The viewpoint of Albert Eberhard Friedrich Schäffle (1831–1903) is stated unambiguously in the title of his book, *The Structure and Life of the Social Body* (1875–1878), and especially in its subtitle, which declares it to be an "encyclopedic sketch of a real anatomy, physiology, and psychology of human society" in which the "national economy [is] considered as the social process of digestion."[193] Schäffle was aware that the correspondence between human society and animal bodies is imperfect since the ties between humans derive from the mind and are not physical. "No uninterrupted occupancy of space," he wrote, is observed in the substance of society, in contrast to the "organic body," in which "cells and intercellular parts form a solid object."[194] That is, in the social body there are no physical forces such as "cohesion, adhesion, or chemical affinity" to "effect coherence and co-ordination," but rather there are "mental forces" which establish "spiritual and bodily connection and co-operation between spatially separated elements."[195] Because of the presence of such statements, Schäffle can be said to have differed from Lilienfeld, as René Worms noted, in that he regarded society only as an "organized" entity, whereas Lilienfeld believed society to be a "fully organic system," an "*organisme concrète*."[196] Yet, as Stark has pointed out, although Schäffle concedes that it is impossible to make a strict comparison between physical and intellectual or psychic entities, nevertheless he insists that there is no "essential difference between social and organic tissues."[197] Accordingly the greater part of Schäffle's four enormous volumes consists of comparisons of the social body and the physiological body.

Like Lilienfeld, Schäffle held that the fundamental unit of society must be the equivalent of the biological cell: his starting point was that "the simplest elements of the bodies of the higher species of plants and animals" are "cells and the intercellular substances interspersed between them."[198] He concluded that "the family has all the traits of the tissue cell," that "every fundamental trait of the structure and function of the organic cell is repeated here."[199] After noting many similarities between organic and social bodies, Schäffle decided that "in all social organs" there is a "tissue which looks after the intake and outflow of the materials of regeneration and nutrition from and to the channels of economic production and circulation and secure[s] a normal digestion on the part of all the elements of the organ or organic part concerned." This "tissue" or social institution is "the household." Schäffle thus made a comparison between the household "and the capillary tissues of the animal body":

> The great social digestive apparatus, in other words, the national economy, the production and circulation of commodities, leads in the end to as many households as the body social has organs and every organ independent tissues and tissue elements.[200]

He found a perfect parallel between the vegetable and animal processes of digestion and the processes of production in human societies, even to the point of believing that "primary production marks the starting point" and "the ejection of human corpses and material waste the end point of external social digestion."[201]

Schäffle found a close resemblance between the streets, roads, buildings, and other constructional elements of the human living space and "the bone and gristle tissue" of animals. As Stark has shown, Schäffle even managed "to find homologues to such protective tissues of the animal body as hair, nails, and horny skin," whose social counterparts (Schäffle himself says they occur "analogically in the body social") are "roofs, coverings, wrappings, fences, walls, clothes, even picture-frames and book-covers. . . ."[202] He also found a likeness between the "collective property [of the social organism] in literature, works of art, roads, transport undertakings, defensive institutions, institutional buildings, and public works equipment" and "the circulating, solvent and protective material which serves the organic body by means of liquidity, softness, elasticity, the chemical counterbalancing of destructive affinities and in other ways." One can hardly overlook, he wrote, the close "analogy of an organic mass of property of the cells in the intercellular materials" and "the social institution of a mass of property in tools, coverings, means of transportation and collective arrangements of all sorts."[203]

René Worms,[204] who devoted a whole treatise to *Organism and Society* (1896), also went beyond merely stating that society is comparable to an organism or is only the analogue of an organism. Like Lilienfeld and Schäffle, he declared that society "constitutes an organism, with something essential in addition." His goal was to go a step beyond Schäffle, "who passes in France as one of the most intransigent partisans of our theory," and Lilienfeld, whom he criticized for having made too many observations "more ingenious than certain," as well as observations "more piquant than decisive." Society, Worms declared, must be an organism because it is a collection of organized living entities fulfilling all the defining requirements of a biological system. He believed that his organismic view of society illustrated Claude Bernard's definition that "vital properties are in reality only in the living cells, all the rest is arrangement and mechanism."[205]

In his detailed development of an organismic sociology, Worms, like his predecessors in this field, introduced extensive biological tutorials and drew heavily on recent work in histology, cellular morphology and physiology, and pathology. A major source was Spencer's *Principles of Sociology*, a work continually cited by him with the highest respect. Proceeding in the manner of a biological analyst, Worms began his treatise with a discussion of the anatomy of societies, then turned to social physiology, and concluded with social pathology.

In his later *Philosophy of the Social Sciences* (1903), Worms confessed that several years after publishing his *Organism and Society* he had been led, "by personal reflection and by discussion," to moderate the "intransigence of my earlier conclusions." Primarily, he admitted to having underestimated the "true value of the individual by making him a simple cell in the social body" and by believing him to be "chained by physio-biological laws." He had accordingly neglected the power of "free will" and the degree to which man is regulated by "the laws which he had given to himself" and the "contracts he made."[206]

In discussing the alterations of his point of view, he gave importance to the ideas of Ernest Haeckel and to the recent discovery of the relative discontinuities among cerebral cells, which he said is "the glory of Golgi and Ramón y Cajal." He also pointed to E. Metchnikoff's research on phagocytes, which had indicated new aspects of cells, even to the extent of suggesting that cells exercise an art or even a science in defending themselves against their enemies. In short, Worms said, his sociological critics had undermined his earlier too-simplistic view of the social organism, while the advances of bio-medical science had been radically altering the scientific base on which the earlier organismic sociology was founded, primarily by enlarging our knowledge and understanding of the life and functions of the cells of which living organisms are made.[207] In the present context, the details of the differences between the two treatises are of less interest than the fact that in each of these works the social parallels drawn from the life sciences were direct reflections of the changing current state of biological and medical knowledge.

In addition to Lilienfeld, Schäffle, and Worms, I should mention the works of Herbert Spencer. (For a fuller discussion of Spencer's use of analogies, see the essay by Victor Hilts in *The Natural Sciences and the Social Sciences*.) Spencer is a challenging figure in the history of ideas. Probably one of the most influential thinkers of the nineteenth century, at least in social thought, he is often disparaged today.[208] For example, the author of a recent article in *Nature*, Jonathan Howard, proposed that "books on the history of evolution" should be "avoided in direct proportion to the number of references they make to Spencer."[209] The ambiguity concerning Spencer is perhaps symbolized by the frequently cited quotation from Charles Darwin's *Autobiography* to the effect that Spencer's *Principles of Biology* made him feel that Spencer was "about a dozen times my superior" and that he believed that Spencer would come to be considered the equal of Descartes and Leibniz. As J. W. Burrow wryly remarked, Darwin rather spoiled the effect by adding "about whom, however, I know very little."[210]

Spencer's extravagant use of analogies has already been noted. He not only employed organic analogies, however, but also would suddenly shift to mechanical analogies, rather than keeping to one or the other. As

Peel observed, Spencer "can speak in terms drawn from physics, and go on: 'changing the illustration and regarding society as an organism.'"[211] For example, noting that "in an animal organism, the soft parts determine the forms of the hard ones," he concluded by analogy that "in the social organism the seemingly fixed framework of laws and institutions is moulded by the seemingly forceless thing – character." This conclusion is soon afterwards restated in an engineering analogy, in which an institution is said to resemble a building because its structure is determined by "the strength of the materials" rather than "the ingenuity of its design."[212]

Toward the end of his life, in his autobiography, Spencer repudiated emphatically "the belief that there is any special analogy between the social organism and the human organism." He concluded:

> Though, in foregoing chapters, comparisons of social structures and functions to structures and functions in the human body, have in many cases been made, they have been made only because structures and functions in the human body furnish the most familiar illustrations of structures and functions in general. . . . Community in the fundamental principles of organization is the only community asserted.[213]

On this score we may agree with Spencer's biographer that "there is a certain incompatibility, if not inconsistency, between Spencer's awareness of the proper logical bounds of the comparison, and the evident pleasure he took in picturesque and striking parallels."[214]

Today's sociological literature exhibits an almost universal disdain for organismic sociology,[215] usually without attempting to find out whether this school of thought, admittedly influential in its own time, has left us a permanent legacy. One important contribution of the organicists has been to transfer to social thinking at large some of the concepts and principles developed in medical science. Comte, Lilienfeld, Schäffle, Worms, and others stressed the medical concepts of normal and pathological. They advocated the important principle (adopted in the first instance by Comte from Broussais) that normal and pathological social states should not be considered wholly different types of conditions but rather extreme stages of a single type of condition. The insistence of these writers on this medical analogy is echoed today in such phrases as "a healthy society." And when the organicists even drew on Virchow's medical pathology and sought for social analogues of the germ theory of disease, they were working in the same mode as those who, in our time, have adopted concepts of psychoanalysis to sociological analysis.

One of Michel Foucault's many prescient though startling conclusions was that only "ignorance" has caused sociologists to seek their origins in Montesquieu and Comte, whereas they should have recognized that "sociological knowledge is formed in practices like those of the doctors."[216] This organicist theme of an analogy between sociology and medicine is a feature of older, historically oriented textbooks such as *An Introduction to the Study of Sociology* by Small and Vincent,[217] and the theme appears explicitly in more recent times, as Bryan Turner found out, in Louis Wirth's notion (in 1931) of "clinical sociology" and "sociological clinics."[218] In 1935 L. J. Henderson put forth the case that sociologists should adopt the analogy of clinical medicine (and even its techniques) and anticipated Foucault in conceiving (in 1936) "the practice of medicine as applied sociology."[219]

To organic sociologists it seemed an obvious analogical conclusion from medicine that social ills or diseases are caused by ailing individuals, just as Virchow taught that medical disorders should be reduced to a pathological condition in individual cells. Even earlier, in the eighteenth century, there was a strong current of thought linking individual health or well-being to the health of society. Utopians such as Condorcet drew an analogy between the eventual achievement of a perfect condition of individual health and the creation of a perfect society, predicting a time when people would become so healthy and long-lived that death would, as he wrote, become a "curious accident."[220] But the studies of Malthus on population took a decidedly different turn and showed that this analogy between health of the individual and the health of society might be too facile. Malthus demonstrated with grim examples that health and natural vigor in procreating could be a cause of social ills and diseases, producing a "power of population" restrained only by misery or vice.[221] The effect of human health and the healthy, natural "desire and power of generation" (as Hume described this human drive[222]) would naturally lead to poverty, hunger, and misery, Malthus argued, because any possible increase in food supply (limited to an arithmetic ratio) could never keep up with the increase of population (in a geometric or exponential ratio). The widespread influence of Malthus and the discussions of the social implications of population studies indicate that biological considerations are not essentially foreign to sociology and may give us a measure of the importance and originality of the organicist sociologist who explored a different and valuable set of analogies.

1.9. INCORRECT SCIENCE, IMPERFECT REPLICATION, AND THE TRANSFORMATION OF SCIENTIFIC IDEAS

In the use of the natural sciences for the advancement of the social sciences, it may happen that the science being applied is simply wrong. A conspicuous example is provided by the American sociologist Carey, who sought to build a science of society on physical principles centering on Newtonian celestial mechanics. I have mentioned Carey's ideas earlier as an example of mismatched homology, in reference to the concept of mass and the form of his law. Carey also made a grave error in stating the law of universal gravity, wrongly believing that the force between two gravitating masses is inversely proportional to the distance between them rather than inversely proportional to the square of the distance between them.[223] Although this error is obvious to anyone who is even slightly familiar with elementary physics, it has not been noted by Carey's critics.[224] Of course, an argument can be made that Carey's system would not have been any better if he had used the correct Newtonian law. Since he did not develop his subject mathematically, his ignorance of the exact form of the law of gravity may be irrelevant. But such a conclusion condemns Carey's sociology for falsely claiming to be based on Newtonian principles. I strongly doubt whether any sociologist – or other social scientist – would advocate that his or her subject be founded on blatantly erroneous science.

In the application of the natural sciences to the social sciences, errors such as Carey's are not so common as misinterpretations and imperfect replications. An example of a misinterpretation appears in Montesquieu's celebrated *Spirit of the Laws* (1748). In discussing the "principle of monarchy", Montesquieu wrote, "It is with this kind of government as with the system of the universe." That is, "there is a power that constantly repels all bodies from the center, and a power of gravitation that attracts them to it."[225] This notion of a "power of gravitation" that "attracts" all bodies to a center is, of course, Newtonian. But Newton's explanation of the "system of the universe" expressly denied any balance of centripetal and centrifugal forces. Montesquieu had only an imperfect understanding of the Newtonian concept of universal gravitation. In the example under consideration, he shows his essential belief in the older framework of Cartesian physics and balanced forces, into which he tried to introduce a quasi-Newtonian concept that does not fit. There is abundant evidence that Montesquieu remained a Cartesian and never fully grasped the principles of the new Newtonian natural philosophy.[226]

An instructive example of a different sort, at first glance seeming to imply an imperfect replication rather than a misunderstanding, occurs in Adam Smith's *Wealth of Nations* (1776), in the discussion of his celebrated concept of "natural price." Smith wrote that the "natural price" is "the central price, to which the prices of all commodities are continually gravitating."[227] The use of the words "all" and "continually gravitating" invoke Newtonian science and may even suggest that this passage is an instance of Smith's alleged Newtonianism in economics. Unlike Montesquieu, Smith had some understanding of Newtonian scientific principles and, in his essay on the history of astronomy,[228] wrote in glowing terms about Newton's scientific achievements.

Smith's use of gravitation in relation to the natural price differs in one important feature from Newtonian or physical gravitation. A basic axiom of Newton's physics is his third law of motion, that action and reaction are always equal. A consequence of this law is that "all" bodies are not only "gravitating" toward some central body, but are also mutually "gravitating" toward one another. The central body, accordingly, must be "gravitating" toward all other bodies in the system. As a result, for Smith's economics to be a complete and accurate replication of Newton's physical theory of gravity, all prices would have to "gravitate" toward one another and the "natural price" would analogically have to "gravitate" to the "prices of all commodities."

Accordingly, we may all the more admire Smith for having only partially replicated the Newtonian physical concept, for having adapted or transformed the Newtonian physical concept in a way that was of use in economics. Only a brash display of historical Whiggism would fault Smith on the grounds of imperfect replication. The fact is that he was creating a concept for the science of economics and not working on a problem in celestial physics, not pursuing research in the applied physics of gravitation.

Smith's use of a gravitating economic force may serve as a reminder that economics is not an exact clone of physics and that the concepts used in economics need not be exact homologues of those originating in physics. This principle has been stated in a most incisive manner by Claude Ménard, who makes the important point that the successful use of analogies is "not simply a transparent transposition of concepts and methods," that the creative use of analogies always "highlights a difference." He concludes that in every "transfer of concepts" from one domain to another "these concepts take on a life of their own in the reorganized science."[229]

Some recent historical studies of economics seem to be based on a different tacit assumption, namely, that a valid social science must not only be an analogue of the natural sciences but must replicate the natural sciences in every degree of homology of concepts and principles. The history of the natural sciences, however, shows that many of the greatest advances have come not so much from a cloning transfer of ideas from one branch of science to another as from a transformation, from a significant alteration of the original. We may see this process in the way in which Newton forged the concept of inertia as a property of mass, the primary step in the revolution that produced modern rational mechanics. The term "inertia" was introduced into physics by Kepler as part of his argument for the Copernican system. In the pre-Copernican systems, such as those of Aristotle and Ptolemy, the earth was stationary, fixed or immobile at the "center" of the world. Thus, in Aristotelian physics, a terrestrial or "heavy" object is said to fall "naturally" toward the earth's center, which is its "natural place," clearly defined and fixed in space at the center of the immobile earth. For a Copernican, however, since the earth is in continual orbital motion, its center has no fixed or permanent place at the center of the world. There is, therefore, no "natural place," in the old Aristotelian sense, for a falling body to seek. Therefore, Kepler postulated that matter is fundamentally "inert" or is characterized by "inertness" or "inertia." Because matter is inert, it cannot move of and by itself but requires a "vix motrix" or "moving force" for motion to occur. If the "moving force" ceases to act, Kepler concluded, the body will necessarily come to rest then and there, wherever it happens to be. He thus eliminated the anti-Copernican dogma of "natural places."

Newton transformed Kepler's idea, keeping the name which Kepler had introduced. That is, he did not replicate Kepler's idea in his own system of physics. In his transformed concept, there was a very different consequence of the body's "inertness" or "inertia." Whenever there is no externally acting force, Newton wrote in Definition Three and in the First Law of Motion in the *Principia* (1687), a body will persevere in either a state of rest of a "state of motion," that is, uniform motion in a straight line. Newton was consciously aware of the difference between his concept and principle of inertia and Kepler's.

In his personal copy of the *Principia*, he entered the comment that he did not mean by "inertia" Kepler's "force of inertia," by "which bodies tend to rest," but rather a "force of remaining in the same state, whether of resting or of moving."[230]

A somewhat similar transformation of scientific ideas was a feature of Darwin's creation of his theory of evolution. At that time, the geologist Charles Lyell interpreted the paleontological record, marked by successive disappearances of species, in terms of a contest for survival among different species. Charles Darwin, contemplating Lyell's ideas while reading in Malthus, transformed Lyell's concept. Darwin had observed that the individual members of any single species differ from one another in heritable characteristics. Recognizing that certain characteristics were better suited than others for survival in a given environment, Darwin made a radical change in Lyell's idea. Rather than supposing a competition for survival among different species, Darwin proposed that the contest takes place among different individuals of the same species, leading over the course of time to species modification. In making this transformation of Lyell's concept, Darwin introduced into biological thought what is known today as "population thinking," which – according to Ernst Mayr – was one of Darwin's most original and most significant innovations.[231]

Such case histories illustrate the enormous force of the human imagination in transforming an existing concept or principle or theory in the natural sciences. These are not "cautionary tales" of erroneous or imperfect replication, but rather illustrations in detail of the creative process in science at its highest degree. Awareness of such case histories, furthermore, may serve to alert the critical historian to a feature that often appears when a natural scientist or a social scientist makes use of concepts, principles, theories, and methods from another domain. Whether the transfer occurs on the level of analogue, homologue, or metaphor, there is commonly some kind of distortion or transformation that arises from the differences between disparate realms of knowledge. Part of the distortion observed by Ménard, Mirowski, and some other critics of neoclassical economics arises from the "absence of laws of conservation in economics." There is, however, no universal agreement among economists that the omission of a conservation principle so distorts the energy analogy that it constitutes an irreparable fault in the foundations of neoclassical economics.[232] For that matter, a number of economists doubt Mirowski's blatant assertion that neoclassical economists "copied their models mostly term for term and symbol for symbol" from physics. For example, in a review of *More Heat than Light* in the *Journal of Economic Literature*,[233] Hal R. Varian disputes Mirowski's "claim that neoclassical economics is 'incoherent' because of the misappropriation of the energy concept." He also rejects Mirowski's parallel claim "that conservation of energy is an inherent aspect of the physical concept of energy, and that this sort of conservation principle is not valid for utility," so that "utility is not an intellectually coherent concept." Varian's conclusion is that Mirowski has only shown "that utility is not energy."[234]

Such questions of distortion or transformation in the transfer of concepts, laws, principles, and theories are different, however, from simple errors of fact. Carey's social law is not the result of a distortion or a creative, non-orthodox interpretation of Newtonian science. Carey simply made an error in physics; he just did not know the correct gravitational law. Similarly, Montesquieu did not distort Newtonian physics, nor did he omit a significant feature (as was the case for Smith and the mutuality of gravitation or of Walras and conservation); rather he misunderstood or did not know the Newtonian explanation of curved orbital motion. I have mentioned that Carey's sociology would not be in any way different if he had known and used the correct Newtonian law. Similarly, Montesquieu's social and political ideas would not be significantly altered by the substitution of a correct for an incorrect Newtonian explanation; it probably would not make much difference to his system or to the thrust of his argument if the Newtonian references were completely eliminated.

There are, however, many examples of fruitful advances in social thought resulting from transfers in which the original concept or principle may not be fully understood. An example is found in the intertwined history of biological and social thought relating to the principle of division of labor, analyzed by Camille Limoges in *The Natural Sciences and the Social Sciences*. Indeed, it is generally known among social scientists that misinterpretations often lead to fruitful results, even when the source is another social science. A celebrated example from political science is the doctrine of the separation of powers, a central feature of the form of government adopted in the Constitution of the United States. One direct source for this principle, as A. Lawrence Lowell has documented, is a misreading of the ideas of Montesquieu.[235]

1.10. INAPPROPRIATE OR USELESS ANALOGIES

All analogies are not equally useful. The extreme case occurs when an analogy is so inappropriate as to have no utility for social science. This is not a matter of personal judgment, but a fact of history. Two analogies that have frequently been used in considering the state or society have proved to be inappropriate. One is taken from the biological or life sciences, the other from the physical sciences. One is part of the organismic analogy of the state as the body politic; the other is the Newtonian analogy of the state or society as a physical system. We have seen how in our own century Walter Cannon conceived that his own researches might give the organismic analogy new life. But Cannon did not provide any significant new insights into the theory of society. Nor have any successors to my knowledge made use of his general analogy in a fruitful way. The only possible conclusion is that, in the form presented, the analogy has proved to be inappropriate for the development of sociological knowledge or understanding. If an analogy does not provide a gauge of the validity of a social theory or system or concept or does not introduce some new insight into the social science, then the analogy, being of no use to the social science, must be deemed inappropriate.[236]

The notion that gravitational cosmology or the Newtonian system of the world could provide an analogy for society or for the ordering of the state goes back to the days of Newton himself. One of his disciplines, Jean-Théophile Desaguliers, author of a standard Newtonian textbook, embodied his hopes in a poem,[237] *The Newtonian System of the World, the Best Model of Government*. No political theorist, no practical politician or political leader, and no natural or social scientist ever made use of this curious presentation. Here then is an example of a useless analogy.

There is another early example of useless or inappropriate analogy that is similarly associated with Newton. It is an attempt by a contemporary of Newton's, the Scots mathematician John Craig, to replicate Newtonian science in human affairs. Craig's *Theologiae Christianae Principia Mathematica* (1699) is a direct emulation of Newton's *Philosophiae Naturalis Principia Mathematica*.[238] Craig's aim was to devise a Newtonian law in a social context in the realm of reliability of testimony. The subject he explored was the degree of credence that may be assigned to the testimony of successive witnesses, a topic of major significance in the context of reported miracles. Craig came up with an ingenious Newtonian answer: the reliability of such testimony varies inversely as the square of the time from that testimony to the present, just as the Newtonian gravitational force decreases as the square of the distance. This law is plainly another example of inappropriate analogy.[239]

Despite the hopes of many social scientists, Newton's physics – i.e., the physics expounded by Newton in the *Principia* – has never provided a useful analogy for economics, political science, or sociology. Although post-Newtonian rational mechanics (with non-Newtonian additions by d'Alembert, Euler, Lagrange, Laplace, and Hamilton) proved useful for economics, especially when combined with energy physics, Newtonian rational mechanics by itself was not sufficient to provide a useful model for the social sciences. The reason, I believe, is that the Newtonian system is built on a set of abstractions and conditions that are not realizable in the world of experience. Even the Newtonian system of the world is an abstract concept to the extent to which it cannot be embodied in a mechanical model or picture, in the sense that is possible for the Cartesian system of vortices or even the complex machinery of the Ptolemaic world of epicycles or the Aristotelian universe of nesting spheres. In fact, it was on account of this feature that some of Newton's contemporaries rejected the celestial physics of the *Principia*, criticizing Newton specifically for having deserted the "mechanical philosophy." In any case, the record of history shows that Newton's physics, despite centuries of hope and effort, has not yielded an analogy appropriate for the social sciences.

Social Darwinism provides another significant illustration of an inappropriate analogy.[240] Whereas evolution in general continues to be useful to the social sciences, social Darwinism has left no permanent scientific legacy. Any strict comparison and contrast of the factors operative in Darwinian biological evolution and those determining success in the struggle to succeed in our modern capitalist society will show at once that Darwinian biological evolution provides an inappropriate analogy for such individual social behavior. In analyzing this example, however, great care must be exercised lest the failure of social Darwinism be seen as a simple example of erroneous science. This would be the case only if social Darwinism had been the result of an application to human society of misunderstood principles of Darwin's theory of evolution. But in social Darwinism it was not Darwinian science that was being applied so much as Spencerian principles.[241] Since a major premise of Spencerian evolution is scientifically incorrect, social

Darwinism does exemplify erroneous science. Yet social Darwinism is not in error for an incorrect interpretation of Darwinian science but rather for adoption of Lamarckian principles of heredity, for rejection of Darwinian evolutionary biology in favor of the biology of Spencerian sociology. This case differs from Carey's error in stating the law of gravity because in Spencer's day, at least until after the discoveries of August Weismann, there was no real proof that acquired characters could not be inherited. Scientifically literate contemporaries of Carey knew the correct law of gravity, while Spencer and his contemporaries could still believe in what later would prove to be incorrect science.

Spencer's evolutionary ideas were scientifically incorrect because he accepted the crudest kind of Lamarckism, believing that a consequence must be that individuals can affect, and even direct, the path of their own evolutionary development. Spencer cannot legitimately be faulted for holding to this belief during most of his career, although he can be criticized for his later quixotic efforts to deny Weismann's research on allegedly scientific grounds, as did his American discipline Lester F. Ward. Furthermore, there are ample grounds for questioning whether there can be any permanent value to Spencerian social theory since it was constructed on a strict Lamarckian basis. Spencer himself admitted his incapability of "separating changes in a group's learning repertory from hereditary modifications."[242]

A somewhat related example of the social application of a scientifically inappropriate analogy, proposed by Stephen Gould, also involves Lamarckian evolution.[243] After examining some aspects of changing technologies, Gould concluded that human culture "has introduced a new style of change to our planet." The reason is that "whatever we learn and improve in our lives, we pass to our offspring as machines and written instructions." Since each generation "can add, improve and pass on," there is a "progressive character to our artifacts," and thus the development of culture may be said to be Lamarckian rather than Darwinian. But "whatever we do by dint of struggle to improve our minds," Gould continues, "confers no genetic advantage upon our offspring," who "must learn these skills from scratch using the tools of cultural transmission."[244] Gould concludes that the "fundamental difference between Lamarckian and Darwinian styles of change" may serve to "explain" why "cultural transformation," unlike biological evolution, is "rapid and linear."[245]

Gould, of course, is not guilty of an error in science. A noted paleontologist and evolutionist, he is explicitly aware that the primary feature of Lamarckian evolution to which he refers in a social context – the inheritance of acquired characters – is not an acceptable principle of natural science. And we have seen him declare that we confer "no genetic advantages upon our offspring" by improving our minds. Hence, he must be arguing that a principle of incorrect natural science may be used to construct a useful principle of social science. But is this Lamarckian analogy really useful?

Gould does not develop any further consequences from his Lamarckian suggestion. He does not explore any alterations that might need to be made in current conceptions of social organization or social change, nor does he even suggest a revision of past and present conceptions of the growth of technology and invention. It was apparently Karl Marx who first suggested that the history of technology should be conceived in a Darwinian mode, but Gould does not mention the fact, nor does he explore in what sense Marx may have used either a Lamarckian or a Darwinian model. Others – most recently George Bassalla – have seen the history of technology as exhibiting a strictly Darwinian framework of evolution, in which the non-Lamarckian principle of natural selection is of primary importance. Their work is also ignored by Gould, who merely suggests a possibly Lamarckian thesis without reference to other versions of evolutionary technology. Hence we may legitimately wonder whether this example is introduced primarily as a metaphor in order to express a point of view about society and technology. In any event, since the analogy is not developed and has not proved to be of use in social analysis, we must as of now assign it to the category of the inappropriate.

Some other analogies in the social sciences, even though based on current and correct natural science, may also prove to be useless or inappropriate and even misleading. They may in the end produce confusion and obfuscation rather than illumination. This aspect of analogy was a central issue in a fairly recent intellectual exchange in economics and will further serve to illustrate a fundamental distinction between analogy and metaphor.[246]

In 1950 Armen A. Alchian published an article on "Uncertainty, Evolution and Economic Theory"[247] which called forth a response by Edith Penrose on the general subject of the use of biological analogies in economics. Penrose admitted, at the outset, that economics "has always drawn heavily on the natural sciences for analogies designed to help in the understanding of economic phenomena."[248] She was not concerned with analogies in general but rather with what she saw as a deleterious effect of using "sweeping analogies" in economics: their tendency to frame "the problems they are designed to illuminate" in so special a way that "significant matters are inadvertently obscured." Concentrating on "theories of the firm," she considered three biological analogies used by economists: the life cycle, natural selection (or viability), and homeostasis.

In the course of her critique, Penrose makes an important distinction between two uses of analogies in economics, a distinction which is similar to the typology which I have been presenting in this chapter. One use is to advance our understanding by referring a not fully understood economic phenomenon to an analogous one in some other science which is presumably better understood. The other, which she calls a "purely metaphorical analogy," uses such resemblances "to add a picturesque note to an otherwise dull analysis" and to help the reader in following a difficult argument or in dealing with a strange concept or principle.[249]

Penrose acknowledges that Alchian's argument is not a crude evolutionism, characterized by value judgments such as beset the social Darwinism of the nineteenth century, but is rather "very modern in its emphasis on uncertainty and statistical probabilities." Among the conclusions on which she focuses her criticism are that "successful innovations – regarded by analogy as 'mutations' – are transmitted by imitation to other firms" and that the "economic counterparts of genetic heredity, mutations, and natural selection are imitation, innovation, and positive profits." She sums up the alleged superiority of the evolutionary analogy "in the claim that it is valid even if men do not know what they are doing." That is, "no matter what men's motives are, the outcome is determined not by the individual participants, but by an environment beyond their control." Thus "natural selection is substituted for purposive profit-maximizing behavior just as in biology natural selection replaced the concept of special creation of species."[250]

Penrose makes an excellent case that on every level of homology (although she does not use this term) there is an incompatibility between biology and economics. For example, she shows that humans differ from other animals in their ability to alter the environment and to become, to some degree, independent of it. Furthermore, she detects a serious error in treating "innovations" as homologues (she writes of "analogues") of "biological mutations," since the latter involve an alteration of the "substance of the hereditary constitution," while innovations rather tend to be "direct attempts by firms to alter their environment." Her conclusion is that the biological analogy has hindered rather than advanced Alchian's stated purpose of exploring "the precise role and nature of purposive behavior in the presence of uncertainty and incomplete information." In effect, Penrose finds that the biological analogy fails on two grounds: it is based on a mismatched homology, and it tends to confuse or obscure rather than to clarify the problem at hand, the economics of the firm. Hence, like the Lamarckian analogy, the analogy of natural selection with respect to economics is inappropriate.

In his reply to the critique, Alchian asserted that his theory of the firm "stands independently of the biological analogy," that "every reference to the biological analogy" was "merely expository" and "designed to clarify the ideas in the theory."[251] In her rejoinder, Penrose reasserted her position that, even so, "the biological analogy places the whole problem in a misleading frame of reference."[252] Wholly apart from the merits of one or the other position with respect to the theory of the firm, Penrose insists that the introduction of the analogy of natural selection hinders rather than furthers understanding. This negative effect of an analogy, even though the analogy is based on correct science, is similar in its net result to that of other varieties of inappropriate or useless analogies: it does not help and may even hinder our understanding.

The problems of specific homologies versus general analogies appears prominently in the revised edition of a book on *Social Change* by the sociologist W. F. Ogburn, of which a major stated aim was to "compare the rate of biological change with the rate of cultural change." In the revised version, published in 1950, Ogburn recalled that the first edition appeared in 1922, at a time when there had been a notable decline in the belief that "the theory of social evolution would explain the origin and development of civilization as the theory of biological evolution had explained the origin and development of man." Darwin, Ogburn noted, "had reduced the evolution of species to three causal factors: variation, natural selection, heredity." Evolutionist theories of society had failed, in Ogburn's opinion, because "many investigators were too slavish in copying the biological account in terms of selection, adaptation, survival of the fittest, variation, survival, recapitulation, and successive stages of development."[253] That is, these theories failed because they adopted a literal homology rather than making use of general analogies or metaphor.

1.11. CONCLUSION

The emulation of the natural sciences by a social science carries with it a validation of the methods used and a legitimation of the enterprise in question. Claude Ménard[254] has expressed this beautifully by referring to the "polemical function of an analogy," explaining that analogy "aims at persuasion, looking to a recognized science for a prestigious answer, for the glamour and security of an argument endorsed by the learned and the revered." An example is the authority carried by a report in the social sciences that has the same formal appearance as one produced in chemistry or physics. In the end, however, the worth of the result will not be gauged by its resemblance to, or even direct kinship with, one or another of the natural sciences so much as by the degree to which it serves its own discipline or by its applicability to the solution of some practical problem.

An allied point is that the use of numerical data, accepted statistical techniques, graphs, and other mathematical tools, including computer modeling, not only makes a social science look like physics but also produces results that are quantitative and testable and hence easily susceptible of application. This is, of course, one of the main reasons why physics *is* an "exact" science and why its results tend to have applications with unambiguous results.

Such considerations are notably illustrated by the Coleman Report, submitted to President Johnson and the Congress in 1966.[255] This appears to have been the first report in the social sciences to originate in a specific mandate from the Congress, embodied in Section 402 of the Civil Rights Act of 1964:

> The Commissioner [of Education] shall conduct a survey and make a report to the President and the Congress, within two years of the enactment of this title, concerning the lack of availability of equal educational opportunities for individuals by reason of race, color, religion, or national origin in public educational institutions at all levels in the United States, its territories and possessions, and the District of Columbia.[256]

Although the actual purpose of the survey was never made explicit, it is obvious in retrospect that one of the questions for which the Congress wanted a documented answer was the relative success of students in integrated and segregated schools. It was plain from the outset that whatever the findings of the survey would be, the whole subject was so controversial that the report would have to be based on the most objective kinds of data possible. Not only did the nature of the inquiry demand that the data be quantitative, but there was the obvious requirement that the collection of data be as free of prejudice as possible and that the statistical analyses be free of any fault in technique. In short, the standards to be adopted were much the same as those that would be used in an investigation in physics or any other of the "exact" sciences.

Of course there were aspects of this study that distinguished it from investigations in the physical sciences. For example, the data collected and used for the Coleman Report were much like census data and therefore less certain than numbers in physics.[257] Again, the choice of factors that were enumerated was not quite so value-free as might have been the case for physics.[258] Furthermore, the Coleman Report had to convince Congress and its constituents of the validity of one of the principal "pathbreaking" findings, that an analysis of "the relation of variation in school facilities to variation in levels of academic achievement" showed that there was "so little relation" that, to all intents and purposes, there was none.[259] The implication was that an increase in financial support of and by itself would not necessarily produce better secondary education. This finding constituted a "powerful critique" of one of the most "unquestioned basic assumptions" or "socially received beliefs" of American education. In support of such consequences, the results of the investigation had to be stated unambiguously in the numerical language of quantitative science.

Ever since the Scientific Revolution, a high value has been set on giving social science the solid foundation of the natural sciences. This goal has traditionally had two very different aspects. One, the subject of this chapter, has been of a limited kind: to make use of the concepts, principles, methods, and techniques of some one of the physical or biological sciences. The other has been greater than merely constructing social theories by introducing analogues or homologues of a particular natural science at a particular time. Adopting the metaphor of the natural sciences traditionally has meant taking on certain features of what was known as the scientific method – supposedly characterized by healthy skepticism, reliance on experiment and critical observation, avoidance of pure speculation, and in particular a specific ladder of steps that would lead (usually by induction) from "facts" to "theory," to a knowledge of the eternal "truths" of nature. This second goal, which might from one point of view seem a more obviously useful aim, has actually become increasingly problematic. Twentieth-century philosophers and historians of science, aided by scientists themselves, have dispelled any belief in "the" scientific method. The extreme position was probably stated by P. W. Bridgman when he declared that insofar as there is any "method" in science it is "doing one's damnedest with one's mind, no holds barred." Accordingly, although many social scientists still aspire to have their subject be "like" the "sciences," the quality of likeness no longer features a specific "scientific method."

Moreover, it is widely recognized today that continuous change (usually characterized as "advance") is a principal feature of the natural sciences. The result is that the particular aspects of any natural science being emulated by social sciences will, often without warning, undergo a radical transformation. Accordingly, the present value or usefulness of principles of social science – just as is the case for principles of social and political practice – can not be reckoned primarily by an evaluative contrast between the present state of some part of physical or biological science and the anterior state current when those principles were being formulated. It is admittedly of general interest and major historical concern to discern whether the economic thought of Adam Smith or of François de Quesnay was in part based on Newtonian or on Cartesian principles of science, but the validity and usefulness of their concepts is not dependent on the present validity of the natural science that originally inspired them. Similarly, the worth of Darwinian evolutionary ideas in sociology or in anthropology has been judged primarily in relation to their use for those social sciences and has not exactly parallelled the ups and downs of scientific consensus on the Darwinian concept of natural selection.

The feature of dramatic change is seen in stark relief in what was long held to be the most paradigmatic of the exact sciences, Newtonian rational mechanics. In the last two centuries, this subject has been altered by the introduction of new principles, such as those associated with d'Alembert, Lagrange, Laplace, and Hamilton, and by the addition of considerations of energy and variational principles; there has been a dramatic and even more radical reconstruction of the whole subject as a result of Einsteinian relativity. And it is much the same in the shift from classical to quantum physics or from the older natural history to molecular biology.

For the historian, the study of interactions between the natural sciences and the social sciences takes on the added dimension of interest because of the feature of change. Historians cannot fail to be impressed when finding that the validity of concepts, principles, laws, and theories in the social sciences transcends the corresponding present validity of the counterparts in the natural sciences that served as the original sources of inspiration or of generation of ideas. This is merely another way of saying that the social sciences have developed an autonomy and do not merely have the status of being instances of applied physical or biological science. This conclusion underlines the importance of the study of history in any study of the methodology, and even of the legitimacy, of the social sciences.

Harvard University

NOTES

* All of the examples introduced in this introductory section are discussed in full, with bibliographic documentation, in the succeeding sections.

[1] On Cournot's important contributions to mathematical economics, see Claude Ménard: *La formation d'une rationalité économique: A.A. Cournot* (Paris: Flammarion, 1978); also Joseph Schumpeter: *History of Economic Analysis* (New York: Oxford University Press, 1954).

[2] See "A Note on 'Social Science' and on 'Natural Science'" in this volume.

[3] On the history of the concept of behavioral sciences, see Bernard Berelson on "Behavioral Sciences" in *International Encyclopedia of the Social Sciences*, ed. David L. Sills, vol. 2 (New York: The Macmillan Company & The Free Press, 1968), pp. 41–45; also Herbert J. Spiro: "Critique of Behavioralism in Political Science," pp. 314–327 of Klaus von Beyme (ed.): *Theory and Politics, Theorie und Politik, Festschrift zum 70. Geburtstag für Carl Joachim Friedrich* (The Hague: Martinus Nijhoff, 1971).

[4] An admirable discussion of the differences in these three usages is given in John Theodore Merz: *A History of European Thought in the Nineteenth Century* (Edinburgh/London: W. Blackwood and Sons, 1903–1914; reprint, New York: Dover Publications, 1965; reprint, Gloucester: Peter Smith, 1976), vol. 1, chs. 1, 2, 3.

[5] See Marie Boas Hall: *All Scientists Now: the Royal Society in the Nineteenth Century* (Cambridge/New York: Cambridge University Press, 1984); also Dorothy Stimpson: *Scientists and Amateurs: A History of the Royal Society* (New York: Henry Schuman, 1948). The Royal Society, however, originally had a large proportion of non-scientists as Fellows, including poets (e.g., John Dryden), doctors, and peers of the realm. After the reorganization of 1847 the non-scientific categories of membership were eliminated, although exceptions are still made (e.g. Prince Philip and the financier-philanthropist Isaac Wolfson).

[6] See n. 21 infra; also Chapter 3 infra.

[7] See Roger Hahn: *The Anatomy of a Scientific Institution: The Paris Academy of Sciences* (Berkeley: University of California Press, 1971).

[8] At first there were four classes: "Physica" (including chemistry, medicine, and other natural sciences), "Mathematica" (including astronomy and mechanics), German philosophy, and literature (especially oriental literature). Later these classes were regrouped into

two major divisions: the natural sciences and mathematics in one and "philosophical and historical" domains in the other. See Erik Amburger (ed.): *Die Mitglieder der Deutschen Akademie der Wissenschaften zu Berlin, 1700–1950* (Berlin: Akademie-Verlag, 1960); Kurt-Reinhard Biermann & Gerhard Dunken (eds.): *Deutsche Akademie der Wissenschaften zu Berlin: Biographischer Index der Mitglieder* (Berlin: Akademie-Verlag, 1960). On the history and vicissitudes of the German Academy, see Werner Hartkopf & Gerhard Dunken: *Von der Brandenburgischen Sozietät der Wissenschaften zur Deutschen Akademie der Wissenschaften zu Berlin* (Berlin: Akademie-Verlag, 1967); the standard history is Adolph Harnack: *Geschichte der Königlich Preussischen Akademie der Wissenschaften zu Berlin*, 3 vols. (Berlin: Reichsdruckerei, 1990).

[9] See note 2 supra.

[10] See Samuel Johnson: *A Dictionary of the English Language*, 2 vols. (London: printed by W. Strahan for J. and P. Knapton, T. and T. Longman, C. Hitch and L. Hawes, A. Millar and R. and J. Dodsley, 1755; photo-reprint, New York: Arno Press, 1979).

[11] *The Works of Lord Macaulay*, ed. Lady Trevelyan, vol. 5 (London: Longmans, Green, and Co., 1871), p. 677. In 1829 Macaulay wrote (ibid, p. 270) that the "noble Science of Politics" was "of all sciences . . . the most important to the welfare of the nations," the science which "most tends to expand and invigorate the mind." Additionally, he declared, the "Science of Politics" is notable among "all sciences" because it "draws nutriment and ornament from every part of philosophy and literature, and dispenses in return nutriment and ornament to all." See also Stefan Collini, Donald Winch, & John Burrow: *That Noble Science of Politics* (Cambridge: Cambridge University Press, 1983), passim, but esp. pp. 102–103, 120. The negative version of Macaulay's statement occurs in the oft-quoted remark made by Bismarck in the Prussian Chamber on 18 December 1863, "Die Politik ist keine exakte Wissenschaft," that is, "Politics is not an exact science." Early in the eighteenth century, in *Gulliver's Travels* (1726), Jonathan Swift regretted that the ignorant Brobdignagians had not as yet "reduced Politicks into a *Science*."

[12] " A Letter to a Noble Lord," in Edmund Burke: *The Works* (London: John C. Nimmo, 1887; reprint, Hildesheim/New York: Georg Olms Verlag, 1975), vol. 5, p. 215; David Hume: *A Treatise of Human Nature*, ed. L.A. Selby-Bigge (Oxford: Clarendon Press, 1896 and reprints), pp. ix and xix–xx. In *An Inquiry Concerning the Human Understanding*, ed. L.A. Selby-Bigge (Oxford: Clarendon Press, 1894), pp. 83–84, Hume wrote of historical "records" as "so many collections of experiments, by which the politician or moral philosopher fixes the principles of his science, in the same manner as the physician or natural philosopher becomes acquainted with the nature of plants, minerals, and other external objects, by the experiments which he forms concerning them."

On Hume and a science of politics, the studies of Duncan Forbes are of primary importance, notably his introduction to the reprint of Hume's *History of Great Britain* (Harmondsworth: Penguin, 1970); "Sceptical Whiggism, Commerce and Liberty," pp. 179–201 of A.S. Skinner & T. Wilson (eds.): *Essays on Adam Smith* (Oxford: Oxford University Press, 1976); "Hume's Science of Politics," pp. 39–50 of G.P. Morice (ed.): *David Hume, Bicentenary Papers* (Edinburgh: Edinburgh University Press, 1977). See also James E. Force & Richard H. Popkin: *Essays on the Context, Nature, and Influence of Isaac Newton's Theology* (Dordrecht/Boston/London: Kluwer Academic Publishers, 1990), ch. 10, "Hume's Interest in Newton and Science" (by J.E. Force).

[13] John Harris, *Lexicon Technicum* (London: printed for Dan. Brown, Tim. Goodwin,

John Walthoe, . . . , Benj. Tooke, Dan. Midwinter, Tho. Leigh, and Francis Coggan, 1704; reprint, New York/London: Johnson Reprint Corporation, 1966 – The Sources of Science, no. 28) defines "Science" as "Knowledge founded upon, or acquir'd by clear, certain and self-evident Principles." Harris says that "Natural Philosophy" is "the same with what is usually call'd *Physicks, viz.* That Science which contemplates the powers of Nature, the properties of Natural Bodies, and their mutual action one upon another."

[14] James S. Coleman: *Foundations of Social Theory* (Cambridge: The Belknap Press of Harvard University Press, 1990), p. xv. On this general topic, see Robert K. Merton: *Social Theory and Social Structure* (Enlarged edition, New York: The Free Press, 1968), ch. 1, "On the History and Systematics of Sociological Theory."

[15] Robert K. Merton: "The Mosaic of the Behavioral Sciences," pp. 247–272 of Bernard Berelson (ed.): *The Behavioral Sciences Today* (New York/London: Basic Books, 1963), esp. p. 256.

[16] A. Hunter Dupree: *Science in the Federal Government: A History of Policies and Activities* (Cambridge: Belknap Press of Harvard University Press, 1957; revised reprint, Baltimore: Johns Hopkins University Press, 1986).

[17] Reported in Henry W. Riecken: "The National Science Foundation and the Social Sciences," *Social Science Research Council Items*, Sept. 1983, **37**(2/3): 39–42, esp. p. 40*a*.

[18] For example, the association of "sociology" with "socialism" is discussed in Albion W. Small & George E. Vincent: *An Introduction to the Study of Society* (New York/Cincinnati/Chicago: American Book Company, 1894), pp. 40–41, where it is stated that "Systematic Socialism has both directly and indirectly promoted the development of Sociology."

[19] Riecken (n. 17 supra), p. 40*b*.

[20] See Chapter 3 infra.

[21] The Ninth Annual Report of the National Science Foundation announced that "during the fiscal year 1959 [in December 1958], the Foundation established an Office of Social Sciences to support research and related activities in the social science disciplines." The Eleventh Annual Report announced that "The Office of Social Sciences was reconstituted as the Division of Social Sciences during fiscal year 1961 [the year ending on 30 June 1961]." The new Division replaced "the previous Social Science Research Program" and was said to represent "a further step in the development of Foundation activities in the area." The social sciences did not maintain this independent importance, however, and there is still some debate on how best to fit the social sciences into the structure of the National Science Foundation.

A note in *Science* (16 Aug. 1991, **253**: 727) on "a proposal to give the social sciences more clout within the agency" summarized the findings of a draft report by a "task force composed of 20 outside social and behavioral scientists and biologists" who recommended the removal of the social and behavioral sciences from their position as part of the Directorate for Biological, Behavioral and Social Sciences. Although most "social scientists" were in favor of this proposal, which would "give them an advocate at the highest level of the agency and win them more funding and respect," some social scientists – it was noted – "don't want to leave the fold." On 23 Oct. 1991, *The Chronicle of Higher Education* (pp. A-23, A-25) reported that the reorganization had occurred, including the establishment of a new Directorate for Social, Behavioral and Economic Sciences. NSF Director Walter E. Massey expressed high hopes that the creation of a "separate new office" would "lead to an increase in funds for those sciences." Although many "social scientists said the change would lead to more recognition and higher budgets," support for the change "was not unanimous" (according to the *Chronicle*'s report) even among the social scientists concerned.

[22] The work of scientists and social scientists mentioned briefly in various sections of this chapter is discussed at greater length and with bibliographical documentation in other sections of this chapter.

[23] Wilhelm Ostwald: *Energetische Grundlagen der Kulturwissenschaften* (Leipzig: Verlag von Dr. Werner Kilinkhardt, 1909). On Ostwald's "Kulturwissenschaften" see Philip Mirowski: *More Heat than Light: Economics as Social Physics, Physics as Nature's Economics* (Cambridge/New York: Cambridge University Press, 1989), pp. 454–57, 132–133, 268.

There were many scientists and social scientists who saw in the new "energetics" a basis for a reconstitution of economics, sociology, history, and so on. A notable example was Henry Adams, who attempted to use J. Willard Gibbs's memoir on the "Equilibrium of Heterogeneous Substances" as the basis of a study on "The Rule of Phase Applied to History." This essay is reprinted along with Adam's "A Letter to American Teachers of History" in Brooks Adams (ed.): *The Degradation of the Democratic Dogma* (New York: The Macmillan Company, 1920).

[24] See Ian Hacking's essay in I. Bernard Cohen (ed.), *The Natural Sciences and the Social Sciences: Some Historical and Critical Perspectives* (Dordrecht: Kluwer, 1994).

[25] See John Brewer: *The Sinews of Power: War, Money and the English State, 1688–1783* (Cambridge: Harvard University Press, 1988), ch. 8, "The Politics of Information, Public Knowledge and Private Interest"; Keith Thomas: "Numeracy in Early Modern England," *Transactions of the Royal Historical Society*, 1987, **37**: 103–132.

[26] See Chapter 2, §3, infra.

[27] See Jacques Roger: *Buffon* (Paris: Fayard, 1989), pp. 234, 296.

[28] Stephen M. Stigler: *The History of Statistics: The Measurement of Uncertainty before 1900* (Cambridge: The Belknap Press of Harvard University Press, 1986), ch. 3, "Inverse Probability." See further, Helen M. Walker: *Studies in the History of Statistical Method* (Baltimore: The Williams & Wilkins Company, 1929; reprint, New York: Arno Press, 1975), pp. 31–38; Hyman Alterman: *Counting People: The Census in History* (New York: Harcourt, Brace & World, 1969).

[29] Keith M. Baker: *Condorcet, from Natural Philosophy to Social Mathematics* (Chicago: University of Chicago Press, 1975), ch. 4.

[30] In addition to Stigler's *History of Statistics* (n. 28 supra), ch. 5, see Theodore M. Porter: *The Rise of Statistical Thinking, 1820–1900* (Princeton: Princeton University Press, 1986), esp. pp. 2, 4; Ian Hacking: *The Taming of Chance* (Cambridge/New York: Cambridge University Press, 1990), chs. 13, 14, 19, 21; Frank N. Hankins: *Adolphe Quetelet as Statistician* (New York: Columbia University Press, 1908; reprint, New York: AMS Press, 1968).

[31] Baker (n. 29 supra), p. 202.

[32] Mirowski (n. 23 supra), ch. 5. Mirowski's thesis has not produced universal acceptance by economists. Not only is it considered extreme, but it is faulted because it does not apply to all founders of neoclassical economics, for example, Karl Menger, and even Léon Walras (see n. 33 infra and §1.5 infra).

[33] See further, §1.6 infra. Walras, we shall have occasion to note, argued for the

similarity of economics and rational mechanics but (see n. 143 infra) did so only after he had produced his system of economics. That is, he did not create his economics by attempting to imitate rational mechanics.

[34] See §1.5 infra.

[35] William Breit & Roger W. Spencer (eds.): *Lives of the Laureates: Seven Nobel Economists* (Cambridge: MIT Press, 1986), p. 74. This topic is developed further in §1.6 infra.

[36] In addition to the works cited in nn. 28, 30 supra, see Gerd Gigerenzer, Zeno Swijtink, Theodore Porter, Lorraine Daston, John Beatty, and Lorenz Krüger: *The Empire of Chance: How Probability Changed Sciences and Everyday Life* (Cambridge/New York: Cambridge University Press, 1989); William Coleman: *Death is a Social Disease: Public Health and Political Economy in Early Industrial France* (Madison: University of Wisconsin Press, 1982).

[37] David Brewster: *Letters on Natural Magic* (New York: Harper & Brothers, 1843), p. 181. In *The Glaciers of the Alps* (Boston: Ticknor & Fields, 1861), p. 285, John Tyndall wrote that "the analogy between a river and glacier moving through a sinuous valley is therefore complete."

[38] Richard Owen defined these two terms as follows: analogue – "A part or organ in one animal which has the same function as another part or organ in a different animal"; homologue – "The same organ in different animals under every variety of form and function." Richard Owen: *On the Archetypes and Homologies of the Vertebrate Skeleton* (London: Richard & John E. Taylor, 1848), p. 7. Despite Owen's phrasing here, the terms "similarity of form" or "sameness of structure" may be used to represent the kind of likeness exemplified by "the same organ." See further n. 40 infra; also Mayr (n. 40 infra), p. 464.

[39] In Darwinian evolution, analogy is the result of parallel adaptation, the way in which different organisms in separate but parallel evolutionary stages have developed, independently of one another, different ways of "adapting themselves to the same external circumstances" or needs. An example is given by an organ of vision, in which a lens concentrates light on special sensitive tissue. Konrad Lorenz has noted that this "invention" had been made independently by animals of four different phyla, in two of which (the vertebrates and the cephalopods) this kind of "eye" has "evolved into the true, image-projecting camera through which we ourselves are able to see the world." See Konrad Z. Lorenz: "Analogy as a Source of Knowledge," *Science*, 1974, **185**: 229–234.

[40] See Ernst Mayr: *The Growth of Biological Thought: Diversity, Evolution, and Inheritance* (Cambridge: The Belknap Press of Harvard University Press, 1982), p. 45, where it is noted that the "term 'homologous' existed already prior to 1859, but it acquired its currently accepted meaning only when Darwin established the theory of common descent. Under this theory the biologically most meaningful definition of 'homologous' is: 'A feature in two or more taxa is homologous when it is derived from the same (or a corresponding) feature of their common ancestor.'"

[41] "Homology" appears with special meanings in several sciences. In addition to the general evolutionary or biological sense, there is the chemical usage (referring to a family of organic compounds in which each member is distinguished from the next in the sequence by some constant factor, notably, a CH_2 group), the mathematical usage (a

topological classification), and a special usage in genetics (to indicate the same linear sequence of genes in two or more chromosomes).

[42] It should be noted that the terms "analogue" and "homologue" are not being used in the present context in the strict sense of evolutionary biology, since in the analysis of the interactions of the social and the natural sciences there is no consideration of "common descent." Furthermore, because "analogy" is often used to designate various types of correspondence, it is sometimes necessary, especially in quoting or paraphrasing the work of others, to employ this term to indicate likeness in more general senses than those specified above.

[43] Henry C. Carey: *Principles of Social Science* (Philadelphia: J. B. Lippincott & Co., 1858), vol. 1, pp. 42–43.

[44] Carey's exact words are: "Gravitation is here, as everywhere else in the material world, in the direct ratio of the mass, and in the inverse one of the distance." In vol. 3, ch. 55, p. 644, Carey recapitulates his physics and social science. He begins by stating "simple laws which govern matter in all its forms, and which are common to physical and social science." The first of these reads: "All particles of matter gravitate towards each other – the attraction being in the direct ratio of the mass, and the inverse one of the distance." Incidentally, it may be observed that Carey has also misunderstood the Newtonian explanation of orbital or curved motion, under the actions of a centripetal force, such as a planet moving under the action of the sun's gravity plus its own component of inertia. Carey says: "All matter is subjected to the action of the centripetal and the centrifugal forces – the one tending to the production of local centres of action, the other to the destruction of such centres, and the production of a great central mass, obedient to but a single law." We may take note that Carey also introduced ratios other than direct and inverse proportion. Thus, in vol. 1, p. 389, he wrote: "The motion of society, and the power of man, tend to increase in a geometrical ratio. . . ."

[45] Although "fallacy" is often used in a narrow technical sense to denote a flaw (or type of flaw) that "vitiates a syllogism," a primary meaning in every dictionary I have consulted (*OED*, *OED*-suppl., *OED* – 2nd ed.; *Concise Oxford Dictionary* – 6th ed.; *Webster's New International* – 2d & 3d eds.) is a misleading argument, or a delusion or error, or some unsoundness or delusiveness or disappointing character of an argument or belief. *The American Heritage Dictionary* gives as the first meaning: "An idea or opinion founded on mistaken logic or perception; a false notion"; other meanings include "the quality of being deceptive" and "incorrectness of reasoning or belief." The only example given is a "romantic fallacy, that Shakespeare was superhuman." This example displays features in common with two frequently encountered uses of "fallacy" today: John Ruskin's notion of the "pathetic fallacy" (in which inanimate objects are supposed to have human emotions) and W.K. Wimsatt and Monroe Beardsley's "intentional fallacy" (overstressing the author's intentions in assessing a literary work). These usages are somewhat similar to Alfred North Whitehead's "fallacy of misplaced concreteness" as presented in *Science and the Modern World* (New York: The Macmillan Company, 1931), ch. 4, pp. 82, 85.

[46] Newton's concept of mass has two separate aspects: one (inertial mass in post-Einstein terminology) is a measure of body's resistance to being accelerated or being made to undergo a change in "state," while the other (gravitational mass) is a measure of a body's response to a given gravitational field (i.e., the weight). For details see my *The Newtonian*

Revolution: with Illustrations of the Transformation of Scientific Ideas (Cambridge/London/New York: Cambridge University Press, 1980).

Newton recognized that in ordinary (i.e., non-relativistic) rational mechanics there is no logical reason why these two concepts or measures of mass should be equivalent. Accordingly, he instituted a series of experiments to show that one is always proportional to the other, that at any given location mass is proportional to weight. These experiments are described in Book 3, prop. 6, of the *Principia*, which reports how he experimented with "gold, silver, lead, glass, sand, common salt, wood, water, and wheat" and could have easily detected a variation of as little as one part in a thousand. Newton, of course, did not use such terms as "gravitational mass" or "inertial mass" but rather proved that for all such materials the ratio of the "weight" to "quantity of matter" (or mass) was the same.

[47] William Jaffé: "Léon Walras's Role in the 'Marginal Revolution' of the Late 1870s," pp. 115–119 of R.D. Collison Black, A.W. Coates, and Craufurd D.W. Goodwin (eds.): *The Marginal Revolution in Economics: Interpretation and Evaluation* (Durham: Duke University Press, 1973).

[48] For a later attempt by Walras to argue that his economics is analogous to Newtonian rational mechanics, see Philip Mirowski & Pamela Cook: "Walras' 'Economics and Mechanics': Translation, Commentary, Context," pp. 189–224 of Warren J. Samuels (ed.): *Economics as Discourse* (Boston/Dordrecht/London: Kluwer, 1990).

[49] Berkeley, for example, produced a very significant critique of the foundations of Newton's theory of fluxions, that is, Newton's version of the calculus. His *Siris* was an attempt "to assimilate Newtonian concepts to the more complex phenomena of chemistry and animal physiology." In his *De motu* he analyzed "Newtonian concepts of gravitational attraction, action and reaction, and motion in general." See Gerd Buchdahl: "Berkeley, George," *Dictionary of Scientific Biography*, vol. 2 (New York: Charles Scribner's Sons, 1970), pp. 16–18.

[50] See §1.9 infra.

[51] For Newton, the motion resulting from an equilibrium of forces can *only* be constant speed along a straight line, not curved motion as along a planetary orbit. Newton's analysis of orbital or curved motion was based on the concept of two independent components. One is an initial component of inertial (or linear) motion, the other a constantly accelerated motion of falling inward toward the center of force. A planet or other orbiting body, of course, does not actually move inward away from the orbit, even though it is constantly falling toward the center; the reason is that the forward motion along the tangent carries that body ahead at such a rate that it continually "falls" away from the tangent to the orbit. Newton said he gave the centrally directed force the name "vis centripeta" in honor of Christiaan Huygens who had made use of the opposite kind of force, "vis centrifuga." For details, see my *Newtonian Revolution* (n. 46 supra). Since orbital motion involves the constant inward (or centrally directed) acceleration of falling, there is no condition of equilibrium.

[52] Berkeley fully understood Newton's explanation. He gave the correct Newtonian reason why the planets do not actually fall inward so as to join together at the center. They "are kept from joining together at the common centre of gravity," he wrote, "by the rectilinear motions the Author of nature hath impressed on each of them." This tangential or linear component, he continued, "concurring with the attractive principle," produces

"their respective orbits round the sun." He concluded that if this linear component of motion should cease, "the general law of gravitation that is now thwarted would show itself by drawing them all into one mass" (George Berkeley: "The Bond of Society," *Works*, ed. A.A. Luce and T.E. Jessop, vol, 7 [London/Edinburgh: Thomas Nelson and Sons, 1955], pp. 226–227).

[53] Ibid., pp. 225–228; cf. George Berkeley: "Moral Attraction," *Works*, ed. Alexander Campbell Fraser, vol. 4 (Oxford: Clarendon Press, 1901), pp. 186–190.

[54] For additional materials concerning Berkeley's Newtonian sociology, see my "Newton and the Social Sciences, with special reference to Economics: The Case of the Missing Paradigm," to appear in Philip Mirowski (ed.): *Markets Read in Tooth and Claw* (Cambridge/New York: Cambridge University Press, 1993 [in press]) – Proceedings of a Symposium at Notre Dame on "Natural Images in Economics," October 1991.

[55] The eminent sociologist Pitirim A. Sorokin translated Berkeley's correct Newtonian physics into a hodgepodge of incorrect pre-Newtonian explanations. Sorokin not only would have Berkeley make use of the misleading notion of a balance of centrifugal and centripetal forces, but continued his travesty by saying that Berkeley concluded that "Society is stable when the centripetal forces are greater than the centrifugal." This is plainly nonsense even in pre-Newtonian physics; if the centripetal forces should be greater than the centrifugal forces, then obviously there would be no stability but an instability, a lack of balance or equilibrium, and a resultant motion inward, as Berkeley clearly stated would be the case under such circumstances. See Pitirim A. Sorokin: *Contemporary Sociological Theories* (New York/London: Harper & Brothers, 1928), p. 11.

[56] See the writings of Duncan Forbes and of James E. Force (n. 12 supra).

[57] David Hume: *A Treatise of Human Nature* (n. 12 supra), pp. 12–13.

[58] If, as Hume believed, human behavior and social action are regulated by social laws, there is implied the possibility of a social science, one in which – as Hume wrote – "consequences almost as general and certain may sometimes be deduced . . . as any which the mathematical sciences afford us." Seeking to establish a kind of psychology of individual action, Hume seems to have envisioned the construction of a new theoretical science that would ultimately find expression in practice. On the certainty of social laws compared to mathematics, see David Hume: "That Politics may be Reduced to a Science," *Essays: Moral, Political, and Literary*, ed. T.H. Green & T.H. Grose (London: Longman, Green and Co., 1882; reprint, Aalen [Germany]: Scientia Verlag, 1964), vol. 1, p. 99.

[59] Cf. *Design for Utopia: Selected Writings of Charles Fourier*, intro. Charles Gide, new foreword by Frank E. Manuel, trans. Julia Franklin (New York: Shocken Books, 1971 [orig. *Selections from the Works of Fourier* (London: Swan Sonnenschein & Co., 1901]), esp. p. 18; *The Utopian Vision of Charles Fourier: Selected Texts on Work, Love, and Passionate Attraction*, trans., ed., intro. Jonathan Beecher and Richard Bienvenu (Boston: Beacon Press, 1971), esp. pp. 1, 8, 10, 81, 84; *Harmonian Man: Selected Writings of Charles Fourier*, ed. Mark Poster, trans. Susan Hanson (Garden City: Doubleday & Company – Anchor Books, 1971). On Fourier, see Nicholas Y. Riasanovsky: *The Teachings of Charles Fourier* (Berkeley/Los Angeles: University of California Press, 1969) and Frank E. Manuel: *The Prophets of Paris* (Cambridge: Harvard University Press, 1962).

[60] It is a fact of record that groups of idealists actually founded Fourierist utopian colonies

along the bizarre lines he suggested and that Fourierism became a considerable political force in several countries.

[61] Emile Durkheim: *The Division of Labor in Society*, trans. George Simpson (New York: The Free Press, 1933; reprint 1964), p. 339. Cf. Durkheim, *De la division du travail social: étude sur l'organisation des sociétés supérieures* (Paris: Félix Alcan, Éditeur, 1893), p. 378; Durkheim, *De la division du travail social*, 5th ed. (Paris: Librairie Félix Alcan, 1926), p. 330. The first and fifth editions are identical at this point.

[62] Ibid., trans., p. 262. In the *Principia* Newton defines a measure of matter which he calls "quantity of matter" (used as a synonym for "body" or "mass") and which he says is proportional to the volume (or "bulk") and density. Durkheim seems to use both volume and mass in the sense of volume; cf., e.g., trans., pp. 262, 266, 268, 339.

[63] Ibid., trans., p. 268. Furthermore (p. 270), the "division of labor is . . . a result of the struggle for existence, but it is a mellowed *dénouement*. Thanks to it, opponents are not obliged to fight to a finish, but can exist one beside the other. Also, in proportion to its development, it furnishes the means of maintenance and survival to a greater number of individuals who, in more homogeneous societies, would be condemned to extinction."

[64] Ibid., trans., pp. 256–282.

[65] Ibid., trans., p. 266. The example was taken, with a direct citation, from Darwin's *Origin of Species*. Darwin, according to Durkheim, found that "in a small area, opened to immigration, and where, consequently, the conflict of individuals must be acute, there is always to be seen a very great diversity in the species inhabiting it. He found turf three feet by four which had been exposed for long years to the same conditions of life nourishing twenty species of plants belonging to eighteen genera and eight classes. This clearly proves how differentiated they are." This was offered in proof of Darwin's observations "that the struggle between two organisms is as active as they are analogous." Since they have "the same needs" and pursue "the same objects," they are rivals. Eventually, as their numbers increase, the resources available no longer suffice for all, and a struggle for survival ensues. But, "if the co-existing individuals are of different species or varieties," they "do not feed in the same manner, and do not lead the same kind of life," and so they "do not disturb each other." What is perhaps most remarkable about Durkheim's argument based on Darwin is the fact that he referred to Darwin at all. It must be kept in mind that at this time, and for many decades afterwards, Darwinian evolution based on natural selection was not regarded with favor by the French scientific establishment.

[66] Durkheim (n. 61 supra, trans.),p. 336; cf. p. 339.

[67] On the uses of organic analogies, see further §1.7 infra. For Theodore Roosevelt, see his *Biological Analogies in History* (New York: Oxford University Press; London: Henry Frowde, 1910); also *Works*, vol. 12 (New York: Charles Scribner's Sons, 1926), pp. 25–60. A. Lawrence Lowell's organismic views of society may be found in numerous works, notably "An Example from the Evidence of History," pp. 119–132 of *Factors Determining Human Behavior* (Cambridge: Harvard University Press, 1937).

[68] Thomas Carlyle: *Sartor Resartus*, introd. H. D. Traill, *The Works of Thomas Carlyle*, 30 vols. (London: Chapman and Hall, 1896–1899; reprint, New York: AMS Press, 1969), vol. 1, p. 172. See Frederick W. Roe: *The Social Philosophy of Carlyle and Ruskin* (New York: Harcourt, Brace & Co., 1921); also David George Hale: *The Body Politic:*

A Political Metaphor in Renaissance English Literature (The Hague/Paris: Mouton, 1971), pp. 134–135.

[69] Thomas Carlyle: *Past and Present* (London: Chapman and Hall, 1843); *Works* (n. 68 supra), vol. 10, p. 137; "Chartism," *Works*, vol. 29, p. 129.

[70] A brief account of Bluntschli's life and career by Carl Brinkmann can be found in *Encyclopaedia of the Social Sciences*, vol. 2 (New York: The Macmillan Co., 1937), p. 606. See also Francis William Coker: *Organismic Theories of the State* (New York: Columbia University; Longmans, Green & Co., Agents; London: P.S. King & Son, 1910 – Studies in History, Economics and Public Law, vol. 38, no. 2, whole n. 101), pp. 104–114. See, further, J.C. Bluntschli: *Denkwürdiges aus meinen Leben*, 3 vols. (Nördlingen: C. H. Beck, 1884); also Friedrich Meili: *J.C. Bluntschli und seine Bedeutung für die moderne Rechtswissenschaft* (Zurich: Drell Füssli, 1908).

[71] Johann Caspar Bluntschli: *Lehre vom modernen Staat*, 6th ed. (Stuttgart: J.G. Cotta, 1885–1886), the first volume of which was translated into English as *Theory of the State* (Oxford: Oxford University Press, 1892); *Psychologische Studien über Staat und Kirche* (Zurich/Frauenfeld: C. Beyel, 1844). Bluntschli was also author of a widely used reference work, *Deutsches Staats-Wörterbuch* (Stuttgart/Leipzig: Expedition des Staats-Wörterbuchs, 1857–1870).

[72] On Rohmer, see Coker (n. 70 supra), pp. 49–60.

[73] Bluntschli's discussion of the sixteen psychological functions of the state was pilloried by Charles E. Merriam: "The Present State of the Study of Politics," *The American Political Science Review*, 1921, **15**: 173–185. Merriam (p. 183) wrote of "Bluntschli's fearfully and wonderfully made 'political psychology,' in which he compared sixteen selected parts of the human body with the same number or organs in the body politic."

[74] *Psychologische Studien über Staat und Kirche* (n. 71 supra), pp. 54, 86–87, cited in translation in Werner Stark: *The Fundamental Forms of Social Thought* (London: Routledge & Kegan Paul, 1962), pp. 61–62.

[75] Ibid.

[76] Ibid.

[77] On Lilienfeld's life and career see §1.8 infra and esp. n. 182 infra.

[78] Paul von Lilienfeld: *La pathologie sociale* (n. 183 infra), p. 59.

[79] Ibid.

[80] Ibid., pp. 59–60.

[81] J. D. Y. Peel: *Herbert Spencer: The Evolution of a Sociologist, (New York: Basic Books, 1971); J.W. Burrow:* Evolution and Society: A Study in Victorian Social Theory (Cambridge: Cambridge University Press, 1970); David Wiltshire: *The Social and Political Thought of Herbert Spencer* (Oxford: Oxford University Press, 1978).

Of a wholly different sort is the analysis of Spencer in Robert J. Richards: *Darwin and the Emergence of Evolutionary Theories of Mind and Behavior* (Chicago/London: The University of Chicago Press, 1987). Richards has made a careful study of Spencer's ideas based on extensive reading and analysis; in particular he has given us a new understanding of Spencer's social views and biological concepts in relation to the main currents of thought in these areas during Spencer's lifetime. For an anti-Spencerian point of view, see Derek Freeman: "The Evolutionary Theories of Charles Darwin and Herbert Spencer," *Current Anthropology*, 1974, **15**: 211–221. See, further, John C. Greene: *Science, Ideology and World View: Essays in the History of Evolutionary Ideas* (Berkeley/Los

Angeles/London: University of California Press, 1981), ch. 4, "Biology and Social Theory in the Nineteenth Century: Auguste Comte and Herbert Spencer"; for a rebuttal, see Ernst Mayr: *Toward a New Philosophy of Biology: Observations of An Evolutionist* (Cambridge/London: The Belknap Press of Harvard University Press, 1988), essay 15, "The Death of Darwin?".

[82] Peel (n. 81 supra), p. 178.

[83] Ibid.; see Herbert Spencer: *Essays: Scientific, Political, and Speculative*, vol. 1 (New York: D. Appleton & Co., 1883), "The Social Organism," pp. 287–289.

[84] See Peel (n. 81 supra), ch. 7 "The Organic Analogy," with comparative examples of Spencer's use of analogies from physics. For the context of Spencer's analogies, see Richards (n. 81 supra).

[85] Spencer (n. 83 supra), pp. 277–279, 283–286.

[86] Herbert Spencer: *Essays Scientific, Political, and Speculative*, vol. 3 (New York: D. Appleton & Co., 1896), "Specialized Administration," pp. 427–428.

[87] René Worms: *Organisme et société* (Paris: V. Giard & W. Brière, 1896), p. 73.

[88] Walter Cannon: "Relations of Biological and Social Homeostasis," pp. 305–324 in his *The Wisdom of the Body* (New York: W. W. Norton & Company, 1932; revised in 1939).

[89] Ibid., pp. 309–310. The significance of the cell theory as a source of analogues for social theory is discussed in §1.8 infra.

[90] Ibid., pp. 312, 314.

[91] See Merton (n. 14 supra), ch. 3, pp. 101n, 102–103.

[92] Walter Cannon: "The Body Physiologic and the Body Politic," Presidential Address to the American Association for the Advancement of Science, in *Science*, 1941, **93**: 1–10.

[93] Ibid.

[94] Henry C. Carey: *The Unity of Law* (Philadelphia: Henry Carey Baird, 1872), pp. 116–127; for a derisive critique see Stark (n. 74 supra), pp. 156–160.

[95] See the discussion of the Newtonian style at the end of §1.5 infra and in my article on "Newton & the Social Sciences," cited in n. 54 supra.

[96] In fact, there are systems of social thought based on models from physics that seem just as ridiculous as those based on biological models, such as Carey's extravagant electrical analogy (n. 94 supra). Another type of extreme model is set forth in Bradford Peck: *The World a Department Store: A Story of Life Under a Cooperative System* (Lewiston [Me.]: B. Peck, c1900).

[97] See the valuable discussion of these topics in Claude Ménard: "La machine et le coeur: essai sur les analogies dans le raisonnement économique," in *Analogie et Connaissance*, vol. 2: *De la poésie à la science* (Paris: Maloine éditeur, 1981 – Séminaires Interdisciplinaires du Collège de France), pp. 137–165; also trans. Pamela Cook & Philip Mirowski as "The Machine and the Heart: An Essay on Analogies in Economic Reasoning," *Social Concept*, December 1988, **5** (no. 1): 81–95. Especially since the translation omits the mathematical appendix and the discussion, it is well to consult the original.

[98] Stark (n. 74 supra), pp. 73–74.

[99] In this connection we may recall once more Whitehead's presentation of the "fallacy of misplaced concreteness"; cf. n. 45 supra.

[100] Works on metaphor include Max Black: *Models and Metaphors: Studies in Language and Philosophy* (Ithaca: Cornell University Press, 1962); Arjo Klamer (ed.): *Conversations with Economists* (Totowa, [N.J.]: Rowman & Lilienfeld, 1983); Donald N. McCloskey: *If You're So Smart: The Narrative of Economic Expertise* (Chicago/London: The University of Chicago Press, 1990); and Andrew Ortony: *Metaphor and Thought* (Cambridge/London/New York: Cambridge University Press, 1979). For a brief but incisive history of the uses of metaphor from antiquity to the present, see Mark Johnson (ed.): *Philosophical Perspectives on Metaphor* (Minneapolis: University of Minnesota Press, 1981), introd.

This topic also appears in discussions of economics, notably in Mirowski (n. 23 supra).

[101] Herbert Spencer: *The Principles of Sociology*, 3rd ed. (New York: D. Appleton and Company, 1897), vol. 1, part 2, §1, "What is a Society?", §2, "A Society is an Organism."

[102] See Schlanger (n. 160 infra).

[103] *Poetics*, 1457*b*, 1459*a*, 148*a*.

[104] That is, Aristotle held that analogy was a special kind of metaphor that involves a four-term ratio. Let the ratio be

evening : *day* :: *old age* : *life*

or

evening is to *day* as *old age* is to *life*

from which we obtain

old age is the evening of life.

Here we have a metaphor in which something (evening) is attributed to something (life) to which it does not belong. The same would be true for

evening is the old age of day.

Jevons (n. 138 infra), p. 627, gives a similar example, based on a prime minister of a state and a captain of a ship, obtaining the relation that a prime minister is captain of the state.

[105] There are a number of works dealing with rhetoric, especially in relation to the science of the seventeenth century, among them David Johnston: *The Rhetoric of Leviathan: Thomas Hobbes and the Politics of Cultural Transformation* (Princeton: Princeton University Press, 1986); Alan G. Gross: *The Rhetoric of Science* (Cambridge/London: Harvard University Press, 1990); Peter Dear (ed.): *The Literary Structure of Scientific Argument* (Philadelphia: University of Pennsylvania Press, 1991); Steven Shapin & Simon Schaffer: *Leviathan and the Air-Pump: Hobbes, Boyle, and the Experimental Life* (Princeton: Princeton University Press, 1985); Marcello Pera: *Scienza e retorica* (Rome/Bari: Laterza, 1991); and M. Pera & William R. Shea (eds.): *Persuading Science: The Art of Scientific Rhetoric* (Canton, [Mass.]: Science History Publications, USA, 1991).

[106] James I: "Speech of 1603," in Charles H. McIlwain (ed.): *The Political Works of James I* (Cambridge: Harvard University Press; London: Humphrey Milford, Oxford University Press, 1918), p. 272; see Hale (n. 68 supra), p. 111.

[107] On the history of the concept of the body politic, see Hale (n. 68 supra).

[108] Ibid.

[109] For James's statement concerning the spleen, see "Speech in Star Chamber, 1616," *Political Works*, p. 343; Hale (n. 68 supra), p. 111, n. 19. See on this subject Marc Bloch: *The Royal Touch; Sacred Monarchy and Scrofula in England and France*, trans. J. E. Anderson (London: Routledge & K. Paul, 1973).
[110] Cf. §1.5 supra and §§1.7 and 1.8 infra.
[111] See the example of Desaguliers in §10 infra. On this subject see Otto Mayr: *Authority, Liberty, & Automatic Machinery in Early Modern Europe* (Baltimore: The Johns Hopkins University Press, 1986); John Herman Randall, Jr.: *The Making of the Modern Mind: A Survey of the Intellectual Background of the Present Age* (Boston: Houghton Mifflin, 1968), ch. 13.
[112] Quoted from Robert S. Hamilton's *Present Status of the Philosophy of Society* (1866) in L. L. Bernard & Jessie Bernard: *Origins of American Sociology: The Social Science Movement in the United States* (New York: Thomas Y. Crowell Company, 1943), p. 711; see p. 265 for a similar quotation concerning "the true PRINCIPIA MATHEMATICA PHILOSOPHIAE SOCIALIS." Hamilton (ibid., p. 258) believed in two sociological principles, one an analogue of the Copernican system, the other an analogue of Newton's law of universal gravity; he did not, however, fully understand Newtonian science and wrote of "centripetal" and centrifugal" forces as balanced "action" and "reaction." It is observed by the Bernards that in this respect Hamilton's law resembles the law of Carey and the law of "cosmic" attraction of Arthur Brisbane. Although Hamilton expressed admiration for Newton, and even held that he himself had propounded Copernican and Newtonian principles of sociology, he also believed that social science might become more nearly an analogue of geology than of sciences such as astronomy, physics, and chemistry. In this regard his opinion was similar to that of R.J. Wright (Bernard & Bernard, p. 306), who held that social science "ought to be compared *not* with . . . Chemistry, or Astronomy, or even Moral Philosophy, or Political Economy; but rather with . . . Geology or Metaphysics."
[113] The Newtonian style is discussed at length in my *Newtonian Revolution* (n. 46 supra) and in my article cited in n. 114 infra.
[114] For a more complete discussion of Malthus's Newtonianism, see my article in Mirowski (n. 54 supra). See, also, Anthony Flew: *Thinking about Social Thinking: the Philosophy of the Social Sciences* (Oxford: Basil Blackwell, 1985), ch. 4, §1.
[115] Thomas Robert Malthus: *An Essay on the Principle of Population as it Affects the Future Improvement of Society* (London: printed for J. Johnson, 1798). This work, published anonymously and often known as the "first essay," is readily available in two reprints, one of which, edited by Antony Flew (Harmondsworth: Penguin Books, 1970), contains also Malthus's *A Summary View of the Principle of Population* (London: John Murray, 1830), which was originally published with the author's name on the title-page. The other, without notes, but with a foreword by Kenneth E. Boulding, is entitled *Population: the First Essay* (Ann Arbor: The University of Michigan Press, 1959). The text of the second edition (1803) was so completely revised and expanded that it is generally considered "almost a new book," sometimes referred to as the "second essay." The text of this version (reprinted from the seventh edition, 1872, but without the appendices) is available as *An Essay on the Principle of Population*, intro. T.H. Hollingsworth (London: J. M. Dent & Sons, 1914 – Everyman's Library). On Malthus, see Thomas Robert Malthus: *An Essay on the Principle of Population – Text, Sources*

and Background, Criticism, ed. Philip Appleman (New York/London: W. W. Norton & Company, 1976 – Norton Critical Editions in the History of Ideas).

[116] See Flew (n. 114 supra).

[117] W. Stanley Jevons: *The Theory of Political Economy*, 2nd ed. (London: Macmillan and Co., 1879), preface; see this preface in later editions, e.g. (New York: Augustus M. Kelley, 1965 – reprint of the fifth edition, 1911), pp. xi–xiv. Jevons was defending himself against the specific charge that in his book "the equations in question continually involve infinitesimal quantities, and yet they are not treated as differential equations usually are, that is integrated" (p. 102). On Jevons's economics, see Margaret Schabas: *A World Ruled by Number: William Stanley Jevons and the Rise of Mathematical Economics* (Princeton: Princeton University Press, 1990).

[118] Mary P. Mack: *Jeremy Bentham: An Odyssey of Ideas, 1748–1792* (London: Heinemann, 1962), p. 264.

[119] This episode and its significance are discussed in I.B. Cohen: *Revolution in Science* (Cambridge, The Belknap Press of Harvard University Press, 1985), suppl. §14.1.

[120] See Shapin & Schaffer (n. 105 supra).

[121] See Mirowski (n. 23 supra); Klamer (n. 100 supra); McCloskey (n. 100 supra).

[122] Mack (n. 118 supra), pp. 275–281.

[123] Explained in Sigmund Freud: "A Note upon the 'Mystic Writing Pad,'" *The Standard Edition of the Complete Psychological Works*, vol. 19 (London: The Hogarth Press, 1961), p. 228. Two decades after *The Interpretation of Dreams*, in *Beyond the Pleasure Principle* (1920), Freud understood more clearly (as he phrased it in 1924) that "the inexplicable phenomenon of consciousness arises in the perceptual system *instead of* permanent traces" (ibid). See also ed. cit., vol. 5, p. 540, and vol. 18, p. 25; in the latter Freud noted further that this distinction had already been made by Breuer.

[124] Ibid. This pad consisted of a resin or plastic plate, covered with two sheets, one of tissue paper and the other of celluloid, on which one could write with a pointed stylus. Lifting the sheets erased the message, but Freud discovered that the erased message could actually be read in the pad's "memory." This mechanical analogue served two functions often found in the use of analogues: (1) to make his earlier hypothetical conjecture seem reasonable enough for him to set forth his ideas in full, and (2) to make his difficult concept of the structure of memory understandable and thus acceptable, to the psychoanalytic community.

[125] Freud: "Civilization and Its Discontents," *Standard Edition* (n. 123 supra), vol. 21, p. 144. See Donald M. Kaplan: "The Psychoanalysis of Art: Some Ends, Some Means," *Journal of the American Psychoanalytic Association*, 1988, **36**: 259–302, esp. 259–260.

[126] Brian Vickers: "Analogy versus Identity: The Rejection of Occult Symbolism, 1580–1680," pp. 95–163 of Brian Vickers (ed.): *Occult and Scientific Mentalities in the Renaissance* (Cambridge: Cambridge University Press, 1984).

[127] Translated by Vickers from Kepler's *Ad Vitellionem Paralipomena* (*Gesammelte Werke*, vol. 1), p. 90.

[128] Letter to Michael Maestlin, 5 March 1605, quoted in Alexandre Koyré: *The Astronomical Revolution: Copernicus – Kepler – Borelli*, trans. R. E. W. Maddison (Paris: Hermann; Ithaca: Cornell University Press, 1973), p. 252 (from Kepler's *Gesammelte Werke*, vol. 15, pp. 171–172).

[129] These "Rules" appeared in all three editions as part of the introduction of Book Three, "On the System of the World," but they were called "Rules" only in the second (1713) and third (1726) editions.

[130] Laplace's *System of the World*, vol. 2, p. 316, as in Jevons (n. 138 infra), p. 638.

[131] Charles Darwin: *The Origin of Species* (London: John Murray, 1859; reprint, Cambridge: Harvard University Press, 1964), ch. 3, p. 63. This is a case of analogy rather than of generalization because it extends a property observed in one group of entities (humans) to other groups of different entities (plants and animals), whereas a generalization extends a property of some members of a given class to other (or, even, to all) members of that class, as in the generalization that all men are mortal.

Darwin drew on the argument from analogy in other parts of the *Origin*. The concept of natural selection was introduced in analogy with man's process of "artificial" selection in breeding pigeons, horses, dogs, and various ornamental and useful plants. A classic use of analogy, as opposed to generalization, occurs in the final chapter of the *Origin*, in Darwin's presentation of the theory of "common descent." He first concluded that all animals had "descended from at most only four or five progenitors, and plants from an equal or lesser number." This led him to remark, "Analogy would lead me one step further, namely, to the belief that all animals and plants have descended from one prototype." He was aware, as he wrote, that "analogy may be a deceitful guide." Yet he found the evidence for common descent to be very persuasive, noting that "all living things have much in common, in their chemical composition, their germinal vesicles, their cellular structure, and their laws of growth and reproduction." This evidence justified his inference "from analogy that probably all the organic beings that have ever lived on this earth have descended from one primordial form, into which life was first breathed."

[132] See Nagel (n. 136 infra), pp. 107–110.

[133] James Clerk Maxwell: "On Faraday's Lines of Force", in W. D. Niven (ed.): *The Scientific Papers of James Clerk Maxwell* (Cambridge: Cambridge University Press, 1890; reprint, New York: Dover Publications, 1965), vol. 1, p. 156.

[134] This was the occasion for Maxwell to make what may be considered the classic statement about the use of what he called "physical analogies" in science. According to Maxwell, "physical analogies" provide a means "to obtain physical ideas without adopting a physical theory." Ernest Nagel (n. 136 infra, p. 109) has explained that Maxwell meant that he could obtain physical ideas without invoking a "theory formulated in terms of some particular model of physical processes." In other words, by "physical analogies" he implied no more than "that partial similarity between the laws of one science and those of another which makes each of them illustrate the other."

[135] On this point, see especially articles and books by Mirowski.

[136] Ernest Nagel: *The Structure of Science: Problems in the Logic of Scientific Explanation* (New York/Burlingame: Harcourt, Brace & World, 1961), pp. 107–117.

[137] See Maxwell (n. 133 supra). See also J. Robert Oppenheimer, "Analogy in Science," *The American Psychologist* 1956, **11**: 127–135, an address to psychologists in which the physicist J. Robert Oppenheimer stated boldly and unequivocally that "analogy is indeed an indispensable and inevitable tool for scientific progress" (p. 129). He at once narrowed the sense of his assertion, trying to make clear what he meant. "I do not mean metaphor," he added, "I do not mean allegory; I do not even mean similarity." Rather, he intended "a special kind of similarity which is the similarity of structure, the simi-

larity of form, a similarity of constellation between two sets of structures, two sets of particulars, that are manifestly very different but have structural parallels."

[138] W. Stanley Jevons: *The Principles of Science: A Treatise on Logic and Scientific Method* (2nd and final edition, reprint, New York: Dover Publications, 1958), p. 631.

[139] Ibid.

[140] Ibid., p. 632.

[141] Jevons (1965; n. 117 supra), p. 102. It was even suggested by Jevons (n. 138 supra, p. 633), on the authority of Lacroix, that "the discovery of the Differential Calculus was mainly due to geometrical analogy, because mathematicians, in attempting to treat algebraically the tangent of a curve, were obliged to entertain the notion of infinitely small quantities." See Schabas (n. 117 supra), pp. 84–88, "Mechanical Analogies."

[142] Jevons (1965; n. 117 supra), p. 105.

[143] Léon Walras: "Economique et mécanique", *Bulletin de Société Vaudoise des Sciences Naturelles*, 1909, **45**: 313–325; Mirowski & Cook (n. 48 supra), pp. 189–224.

Francis Ysidro Edgeworth proposed the same kind of analogy between his "mathematical psychics" (as he called his brand of economics) and mathematical physics, declaring that "every psychical phenomenon is the concomitant, and in some sense the other side of a physical phenomenon." He had no doubt that "'Mécanique Sociale' may one day take her place along with 'Mécanique Céleste,' throned each upon the double-sided height of one maximum principle, the supreme pinnacle of moral as of physical science." See F. Y. Edgeworth's *Mathematical Psychics: An Essay on the Application of Mathematics to the Moral Sciences* (London: C. Kegan & Co., 1881), esp. pp. 9, 12.

[144] See Claude Ménard (n. 97 supra).

[145] Vilfredo Pareto: "On the Economic Phenomenon: A Reply to Benedetto Croce," translated from Italian by F. Priuli in Alan Peacock, Ralph Turvey, & Elizabeth Henderson (eds.): *International Economic Papers*, vol. 3 (London: Macmillan and Company, 1953), p. 185. For a discussion of Pareto's point of view see Mirowski (n. 23 supra), pp. 221–222; also Bruna Ingrao: "Physics and Pareto's Economics," to be published in Mirowski (n. 54 supra).

[146] Vilfredo Pareto: *Sociological Writings*, ed. S. E. Finer, trans. Derick Mirfin (Oxford: Basil Blackwell, 1966), pp. 103–105, selected from Pareto's *Cours d'économic politique* (Lausanne, 1898), vol. 2, §§580, 588–590.

[147] Ibid. See also Bruna Ingrao: "L'analogia meccanica nel pensiero di Pareto," in G. Busino (ed.), *Pareto oggi* (Bologna: Il Mulino, 1991); and her chapter cited in n. 145 supra.

[148] J.E. Cairnes: *The Character and Logical Method of Political Economy* (New York: Harper & Bros., 1875) p. 69. See Mirowski (n. 23 supra), p. 198.

[149] Leonard Huxley (ed.): *Life and Letters of Thomas Henry Huxley*, vol. 1 (London: Macmillan, 1900), p. 218.

[150] See Mirowski (n. 23 supra), pp. 218–219, 287; William Stanley Jevons: *The Principles of Economics* (London: Macmillan and Co., 1905), p. 50; Jevons: *The Theory of Political Economy* (1965; n. 117 supra), pp. 61–69; Jevons: *The Principles of Science* (n. 138 supra), pp. 325–328; Jevons: *Papers and Correspondence of William Stanley Jevons*, vol. 7, ed. R.D. Collison Black (London: Macmillan, in association with the Royal Economic Society, 1981), p. 80.

[151] Léon Walras: *Elements of Pure Economics*, trans. William Jaffé (Homewood [Ill.]:

Richard D. Irwin; London: George Allen & Unwin; reprint, Philadelphia: Orion Editions, 1984), Preface to the fourth edition, pp. 47–48; also p. 71.

[152] Ménard (n. 97 supra).

[153] Albert Jolink: "'Procrustean Beds and All That': The Irrelevance of Walras for a Mirowski-Thesis," to appear in 1993 in a special issue of *History of Political Economy*, edited by Neil de Marchi, containing papers presented at a symposium (held at Duke University in April 1991) on Mirowski's *More Heat than Light.*

[154] Pareto (n. 146 supra), *Sociological Writings*, p. 104; *Cours*, vol. 2, §592; see the article by Bruna Ingrao, cited in n. 145 supra.

[155] Mirowski (n. 23 supra), pp. 222–231. Extracts from this manuscript, preserved in the Sterling Library, Yale University, are quoted by Mirowski (pp. 228–229, 409 n. 5).

[156] See, e.g., Hal Varian's review of Mirowski's *More Heat than Light* in the *Journal of Economic Literature*, 1991, **29**: 595–596.

[157] This was part of an article on "Distribution and Exchange" in the *Economic Journal* for March 1898 and reprinted in A. C. Pigou (ed.): *Memorials of Alfred Marshall* (London: Macmillan and Co., 1925), pp. 312–318.

[158] Marshall was repeating here the sentiments he had expressed in his inaugural lecture as professor of economics at Cambridge University, printed in Pigou (n. 157 supra; see §1.8 infra).

[159] I have chosen these three organismic sociologists – one Russian, one Austrian, and one French – because their writings exemplify the main issues in the interactions of the natural and the social sciences. There are many others whose writings show the same features, notably the Germany biologist Oscar Hertwig and the Italian sociologist Corrado Gini.

[160] On organismic sociology see F. W. Coker: "Organismic Theories of the State: Nineteenth Century Interpretations of the State as Organism or as Person," *Studies in History, Economics and Public Law* (New York: Columbia University, 1910), vol. 38, no. 2, whole number 101; Ludovic Gumplowicz: *Geschichte der Staatstheorien* (Innsbruck: Universitäts-Verlag Wagner, 1926); Sorokin (n. 55 supra), ch. 4, "Biological Interpretation of Social Phenomena"; Werner Stark (n. 74 supra), part 1, "Society as an Organism"; Judith Schlanger: *Les métaphores de l'organisme* (Paris: Librairie Philosophique J. Vrin, 1971).

Some further major secondary sources on the subject of organismic sociology are: Arnold Ith: *Die menschliche Gesellschaft als sozialer Organismus: Die Grundlinien der Gesellschaftslehre Albert Schäffles* (Zurich/Leipzig: Verlag von Speidel & Wurzel, 1927); Niklas Luhmann: *Die Wirtschaft der Gesellschaft* (Frankfurt: Suhrkamp, 1988); N. Luhmann: *Die Wissenschaft der Gesellschaft* (Frankfurt: Suhrkamp, 1990); D. C. Phillips: "Organicism in the Late 19th and Early 20th Centuries," *Journal of the History of Ideas*, 1970, **31**: 413–432; E. Scheerer: "Organismus," pp. 1330–1358 of J. Ritter (ed.): *Historisches Wörterbuch der Philosophie* (Darmstadt: Wissenschaftliche Verlagsgesellschaft, 1971).

Still valuable as sources of information are certain older works, notably Ezra Thayer Towne: *Die Auffassung der Gesellschaft als Organismus, ihre Entwicklung und ihre Modifikationen* (Halle: Hofbuchdruckerei von C. A. Kammerer & Co., 1903); Erich Kaufmann: *Über den Begriff des Organismus in der Staatslehre des 19. Jahrhunderts* (Heidelberg: C. Winter, 1908).

None of these works, however, pays any attention to the specific relation of these nineteenth-century organismic social scientists to the currents of discovery in the life sciences in their own time.

[161] Alfred Marshall: *The Present Position of Economics: An Inaugural Lecture Given in the Senate State House at Cambridge, 24 February, 1885* (London: Macmillan and Co., 1885), pp. 12–14; this lecture is reprinted in Pigou (n. 157 supra), pp. 152–174.

[162] As most people are aware (because of the interest which Sigmund Freud and Josef Breuer had in this subject), hysteria was a major focus of psychiatric attention in the nineteenth century. An example of hysteria has been introduced in §1.4.

[163] As explained in the Preface to this volume, there is no attempt to discuss all aspects of biological science that have interacted with the natural sciences. I have not dealt with the subject of Darwinian evolution because this interaction is far too complex to be considered in a summary fashion and because it is already the subject of a vast literature that is a continuing part of the current Darwin "industry." Some major aspects of this subject, with special reference to America, are developed in an important way in Robert Richards's *Darwin and the Emergence of Evolutionary Theories of Mind and Behavior* (n. 81 supra), a work that can be especially commended for its methodological approach. Among recent contributions to this general area are Carl N. Degler: *In Search of Human Nature: The Decline and Revival of Darwinism in American Social Thought* (New York/Oxford: Oxford University Press, 1991), and Dorothy Ross: *The Origins of American Social Thought* (Cambridge/New York: Cambridge University Press, 1991). Also worthy of mention is Cynthia Eagle Russett: *Darwin in America: The Intellectual Response, 1865–1912* (San Francisco: W. H. Freeman, 1976).

[164] Auguste Comte: *The Foundations of Sociology*, ed. Kenneth Thompson (New York: John Wiley & Sons, 1975), p. 142. The text is taken from the English translation of Auguste Comte's *System of Positive Polity*, 4 vols (London: Longmans Green, 1877), translated by a group of scholars from *Système de politique positive* (Paris, 1848–1854), vol. 2, pp. 367–382.

Comte believed that Broussais's principle of continuity was especially important in considering the "opposite" mental states of "reason and madness." If the mind surrendered itself to the sense impressions of the external world "with no due effort of the mind within," the result would be "pure idiocy." Madness, in all its intermediate degrees results from the relative failure of the "apparatus of meditation" to "correct the suggestions made by the apparatus of observation." This phenomenon could, he asserted, be studied better in Cervantes's *Don Quixote* "than in any treatise of biology." It could also be traced to "the great principle of Broussais" and could then be "applied to society" as Comte had "now done for the first time." See Comte's *Cours de philosophie positive* (Paris, 1830–1842), quoted in Gertrud Lenzer (ed.): *Auguste Comte and Positivism: The Essential Writings* (New York/Evanston/San Francisco: Harper & Row, 1975), p. 191, taken from *The Positive Philosophy of Auguste Comte*, trans. [and condensed by] Harriet Martineau (London: Longmans, Green, 1853), book 5, ch. 6.

[165] In Lenzer (n. 164 supra), p. 191.

[166] See Edmund Beecher Wilson: *The Cell in Development and Heredity* (New York: The Macmillan Company, 1896; reprint of 3rd ed., New York: The Macmillan Company, 1934), esp. pp. 1–2. Although preliminary steps can be traced to earlier scientists, it was not until the 1840s – largely as a result of the work of J.M. Schleiden and espe-

cially Theodor Schwann – that biologists generally began to give cell theory full serious consideration.

[167] In nineteenth-century thought the principle of division of labor was usually credited to Adam Smith, who displayed it in a dramatic fashion in the opening pages of *The Wealth of Nations*, even though there were other contenders for the invention, including both Benjamin Franklin and Sir William Petty.

[168] This difference is discussed by all organicist sociologists, e.g., Spencer, Lilienfeld, Schäffle.

[169] René Worms called attention to two limitations of this analogy which had been stressed by Herbert Spencer. The first is that, although each individual in the social organism has consciousness, in the animal organism only the organism as a whole, and not the individual cell, has this property. The second: in the social organism the purpose of society, or the organism as a whole is to sustain the lives of the individuals, whereas in the animal or plant the lives of the individual cells serve to support the life of the organism as a whole. Despite these dissimilarities, the cell theory seemed to provide nature's own model on the microscopic scale for the study of human societies, much as the social behavior of ants has done in our own days.

[170] As in societies, the development of the embryo produces special cells and groups of cells with forms and structures suited to their function. This concept of "division of labor," as we have seen (n. 61 supra), originated in social science, then was transferred to the life science and finally migrated back to the social sciences. This transfer is the subject of chapter 10 infra.

[171] On von Baer see the article by Jane Oppenheimer in the *Dictionary of Scientific Biography*, vol. 1 (New York: Charles Scribner's Son, 1970), pp. 385–389.

[172] See Steven J. Gould: *Ontogeny and Phylogeny* (Cambridge: The Belknap Press of Harvard University Press, 1977).

[173] In this section I have not dealt particularly with Herbert Spencer, although he is perhaps the most important of all the organicist social scientists. One reason is that, unlike many other organic sociologists, he did not concentrate attention on bio-medical discoveries relating to the cell theory, although he did make use of cell biology in his writings on sociology. Some of Spencer's uses of biological science in relation to sociology are discussed in §1.4 supra and §1.8 infra. On the subject of Spencer and sociology, see Richards (n. 81 supra). Cf. also n. 208 supra.

[174] Herbert Spencer: *First Principles* (London: Williams and Norgate, 1862), §119. Spencer evidently learned this law from William Carpenter; see Richards (n. 81 supra), p. 269. Richards observes that Carpenter thought that von Baer's law (that "a heterogeneous structure arises out of one more homogeneous") had great generality. Carpenter wrote that "if we watch the progress of evolution [i.e., embryonic development], we may trace a correspondence between that of the germ in its advance towards maturity, and that exhibited by the permanent condition of the races occupying different parts of the ascending scale of creation."

[175] Herbert Spencer, "Reasons for Dissenting from the Philosophy of M. Comte," *Essays: Scientific, Political, and Speculative*, vol. 2 (New York: D. Appleton and Company, 1896), pp. 118–144.

[176] Ibid., pp. 137–138. It should be noted also that cellular embryology reinforced another

principle of organismic sociology. Embryologists revealed that as an individual organism progresses through more and more complex forms, the component cells exhibit structures suitably adapted for their special function, that is, they show the form necessary for the "division of labor." Spencer held that even before encountering von Baer's "law," he had begun to conceive of both "the development of an individual organism and the development of the social organism" as an advance from "independent parts to mutually-dependent unlike parts – a parallelism implied by Milne-Edwards' doctrine of the 'physiological division of labor.'"

[177] For Virchow, the concept of the "cell state" was particularly significant because there was always a close parallel between his "biological views and his liberal political opinions." See Owsei Temkin: "Metaphors of Human Biology," in Robert C. Stauffer (ed.): *Science and Civilization* (Madison: University of Wisconsin Press, 1949), p. 172. Temkin is summarizing Ernst Hirschfeld, "Virchow," *Kyklos: Jahrbuch des Instituts für Geschichte der Medizin an der Universität Leipzig*, 1929, **2**: 106–116. See also Erwin H. Acherknecht: *Rudolf Virchow: Doctor, Statesman, Anthropologist* (Madison: The University of Wisconsin Press, 1953); reprint (New York: Arno Press, 1981).

[178] Rudolf Virchow: *Cellular Pathology As Based upon Physiological and Pathological Histology*, trans. Frank Chance (New York: Robert M. DeWitt, 1860), p. 40; also (London: John Churchill), pp. 13–14.

[179] Virchow, we may note, was not the only nineteenth-century biologist to use social analogies in scientific discourse. Thomas Henry Huxley made use of a social analogy in describing the sponge, which – he said – represented a kind of sub-aqueous city, "in which the people are arranged about the streets and roads, in such a manner, that each can easily appropriate his food from the water as its passes along." This is an example of the use of analogy to illustrate a scientific concept, making such a concept easier to visualize or to understand.

[180] See Temkin (n. 177 supra), p. 175.

[181] Ibid.

[182] Paul von Lilienfeld, or Paul de Lilienfeld, or Pavel Fedorovich Lilienfeld-Toailles, or Pavel Fedorovich Lilienfel'd Toal' (1829–1903), was a Russian civil servant who held responsible government posts and whose avocation was sociology. He published a book in Russian on the elements of political economy in 1860 under the pseudonym "Lileyewa." Another work, appearing first in 1872 in Russian, under the initials P. L., bore the title, *Thoughts on the Social Science of the Future*, which was expanded into a five-volume German version, *Gedanken über die Socialwissenschaft der Zukunft* (vols. 1–4: Mitau: E. Behre's Verlag, 1873–1879; vol. 5: Hamburg: Gebr. Behre's Verlag; Mitan: E. Behre's Verlag, 1881). Of particular importance are *La pathologie sociale* (Paris: V. Giard & E. Brière, 1896) and *Zur Vertheidigung der organischen Methode in der Sociologie* (Berlin: Druck und Verlag von Georg Reimer, 1898). In 1897–1898 Lilienfeld was president of the Institut International de Sociologie. See Otto Henne am Rhyn: *Paul von Lilienfeld* (Gdansk, Leipzig, Vienna: Carl Hinstorff's Verlagsbuchhandlung [n.d.] – Deutsche Denker und ihre Geistesschöpfungen, ed. Adolf Hinrichsen, vol. 6). For further bibliography related to Lilienfeld, see Howard Becker: "Lilienfeld-Toailles, Pavel Fedorovich," *Encyclopaedia of the Social Sciences*, vol. 9 (New York: The Macmillan Company, 1933, 1937), p. 474. See also n. 160 supra for a list of publications relating to organismic sociology. (The first four volumes of the *Gedanken* in one of the sets in

the Harvard University Library contain book plates indicating that they were "bought with the income from the bequest of James Walker . . . former president of Harvard College; 'preference being given to works in the intellectual and moral sciences.' ")

[183] Trans. from *Gedanken*, vol. 1, p. v.

[184] Trans. from *Pathologie*, p. xxii.

[185] Ibid.

[186] Ibid., p. 8.

[187] Ibid., pp. 8–11. Lilienfeld was noted in his own time for his discussion of social diseases that were analogues of diseases of the nervous system, particularly psychological disorders. We have seen (§1.4 supra) an example of his suggestion of a parallel between medical and social disorders in the social analogue of the intellectual and moral state of women suffering from hysteria.

[188] Ibid., pp. 20–21.

[189] Ibid., p. 21.

[190] Ibid., p. 24.

[191] Ibid., pp. 46–47.

[192] Ibid., p. 307.

[193] *Bau und Leben des socialen Körpers: Encyclopädischer Entwurf einer realen Anatomie, Physiologie und Psychologie der menschlichen Gesellschaft mit besonderer Rücksicht auf die Volkswirthschaft als socialen Stoffwechsel*, 4 vols. (Tübingen: H. Laupp'sche Buchhandlung, 1875–1878).

Albert Eberhard Friedrich Schäffle (1831–1903), a German sociologist and economist, was a professor at the University of Tübingen, later moving to the University of Vienna. He was, for a while, a member of the Austrian cabinet. He edited a journal entitled *Zeitschift für die Gesamte Staatswissenschaft*. He envisioned a "rational social state," a kind of utopian blend of capitalism and socialism. He was known in his own times primarily for his exposition of organismic social theory, especially his use of specific biological analogies. See the article on him by Fritz Karl Mann in *Encyclopaedia of the Social Sciences*, ed. Edwin R.A. Seligman (New York: The Macmillan Company, 1934), vol. 13, pp. 562–563. There is no biography of Schäffle in the more recent *International Encyclopedia of the Social Sciences*. See Arnold Ith (n. 160 supra) and Stark (n. 74 supra), pp. 62–72.

[194] Schäffle (n. 193 supra) vol. 1, p. 286; see Stark (n. 74 supra), p. 63. The extracts from Schäffle are quoted from Stark's translation.

[195] Schäffle, vol. 1, p. 286; Stark, pp. 63–64.

[196] Preface to Lilienfeld's *La pathologie sociale* (n. 182 supra), p. vii; cf. Stark, p. 63.

[197] Schäffle, vol. 1, p. 286; Stark, p. 64.

[198] Schäffle, vol. 1, p. 33; Stark, p. 66.

[199] Schäffle, vol. 1, p. 57; Stark, p. 67.

[200] Schäffle, vol. 1, p. 324; Stark, p. 67.

[201] Schäffle, vol. 1, p. 335; Stark, p. 67.

[202] Schäffle, vol. 1, pp. 327, 329; Stark, p. 68.

[203] Schäffle, vol. 1, p. 94; Stark, p. 68.

[204] René Worms (1869–1926), a French sociologist, was educated at the Ecole Normale Supérieure. In 1893 he founded both the Paris-based Institut International de Sociologie and the *Revue Internationale de Sociologie*. He also founded and edited a series of fifty

books on sociological subjects by authors from many countries. He was known in his lifetime particularly for his views concerning the interrelations among "the three disciplines of psychology, social psychology, and sociology." See the biography and critical analysis by Terry N. Clark in *International Encyclopedia of the Social Sciences*, ed. David L. Sills, vol. 16, pp. 579–581 (New York: The Macmillan Company & The Free Press, 1968).

An account of the life and career of René Worms may be found in an article by V.D. Sewny in the *Encyclopaedia of the Social Sciences*, vol. 15 (New York: The Macmillan Company, 1934), pp. 498–499. See also Stark (n. 74 supra).

[205] Worms (n. 87 supra), p. 43.

[206] René Worms: *Philosophie des sciences sociales* (Paris: V. Giard & E. Brière, 1903), vol. 1, p. 53.

[207] Ibid., chs. 2, 3.

[208] See, especially, Derek Freeman, "The Evolutionary Theories of Charles Darwin and Herbert Spencer," *Current Anthropology*, 1974, **15**: 211–237. Cf. also n. 173 supra.

[209] *Nature*, 1982, **296**: 686–687.

[210] J.W. Burrow: *Evolution and Society: A Study in Victorian Social Theory* (Cambridge: The University Of Cambridge Press, 1970), p. 182.

[211] Peel (n. 81 supra), p., 174, including a quotation from Spencer's *Social Statics*.

[212] The organic metaphors predominate in many essays (notably "The Social Organism" [1860]) and in his books, especially *Social Statics* (1850), *The Study of Sociology* (1873), and *The Principles of Sociology* (1876). See Peel (n. 81 supra), ch. 7, esp. p. 174.

[213] Quoted in Peel (n. 81 supra), p. 179.

[214] Ibid., p. 178.

[215] I have not felt the need to make a parade here of the mismatched homologies that appear in the writings of Lilienfeld, Schäffle, Worms, and Spencer (see §1.4 supra), because my aim has been to examine the historical use of analogies rather than merely to call attention to their extravances (as has been done in §1.4 supra).

[216] Michel Foucault: *Power/Knowledge: Selected Interviews and Other Writings, 1972–1977*, ed. Colin Gordon (Brighton: Harvester Press, 1980), p. 151.

[217] See Small and Vincent (n. 18 supra).

[218] Bryan S. Turner: *The Body and Society: Explorations in Social Theory* (Oxford: Basil Blackwell, 1984), pp. 49–50; Louis Wirth: "Clinical Sociology," *American Journal of Sociology*, 1931, **37**: 49–66.

[219] L.J. Henderson: "Physician and Patient as a Social System," *New England Journal of Medicine*, 1935, **51**: 819–823; "The Practice of Medicine as Applied Sociology," *Transactions of the Association of American Physicians*, 1936, **51**: 8–15. These and other papers of Henderson on similar subjects have been edited with an important introductory statement by Bernard Barber: *L.J. Henderson on the Social System: Selected Writings* (Chicago: University of Chicago Press, 1970).

See, on this subject, Talcott Parsons: *The Social System* (Glencoe, Ill.: Free Press, 1951) and "The Sick Role and the Role of the Physician Reconsidered," *Milbank Memorial Fund Quarterly*, 1975, **53**: 257–278.

[220] Marie-Jean-Antoine-Nicolas Caritat, Marquis de Concorcet: *Esquisse d'un tableau historique des progrès de l'esprit humain* (Paris: Agasse, 1795); also Baker (n. 29 supra), pp. 348–349, 368–369.

[221] In later editions Malthus, attempting to lessen the gloomy prospect he had set forth, introduced the power of "moral restraint" as a factor in population control. See n. 115 supra.

[222] David Hume: "Of the Populousness of Ancient Nations," vol. 1 of his *Essays, Moral, Political, and Literary* (Edinburgh: R. Fleming and A. Alison for A. Kincaid, 1742), p. 376. See Catherine Gallagher: "The Body versus the Social Body," pp. 83–106 of Catherine Gallagher & Thomas Laqueur (eds.): *The Making of the Modern Body: Sexuality and Society in the Nineteenth Century* (Berkeley/Los Angeles: University of California Press, 1987).

[223] Carey (n. 43 supra); see §1.4 supra.

[224] See Sorokin (n. 55 supra) and Stark (n. 74 supra).

[225] *The Spirit of the Laws*, trans. Thomas Nugent (revised ed., London: George Bell and Sons, 1878; reprint, New York: Hafner Press, 1949), bk. 3, §7, "The Principle of Monarchy."

[226] On this score see Henry Guerlac: "Three Eighteenth-Century Social Philosophers: Scientific Influences on their Thought," *Daedalus*, 1958, **87**: 6–24; reprinted in Henry Guerlac: *Essays and Papers in the History of Modern Science* (Baltimore: The Johns Hopkins University Press, 1977), pp. 451–464.

[227] Adam Smith: *An Inquiry into the Nature and Causes of the Wealth of Nations* (Oxford: Oxford University Press, 1976 – The Glasgow Edition of the Works and Correspondence of Adam Smith, II), bk. 1, ch. 7, p. 15 (§15). The "Cannan edition" – Adam Smith: *An Inquiry into the Causes of the Wealth of Nations*, ed. Edwin Cannan (London: Methuen & Co., 1904; reprint, Chicago: The University of Chicago Press, 1976; reprint New York: Modern Library, 1985) – is easier to read and has the advantage of useful postils. A postil (1976 ed., p. 65; 1985 ed., p. 59) repeats the message: "Natural price is the central price to which actual prices gravitate."

[228] Adam Smith: *Essays on Philosophical Subjects*, ed. W.P.D. Wightman & J.C. Bryce (Oxford: Oxford University Press, 1980 – The Glasgow Edition of the Works and Correspondence of Adam Smith), vol. 3, pp. 33–105, "The History of Astronomy."

[229] Ménard (n. 97 supra; 1988).

[230] For details see my *Introduction to Newton's 'Principia'* (Cambridge: Harvard University Press; Cambridge: Cambridge University Press, 1971), ch. 2, §1. Newton referred to the inertial property of bodies as both a "vis inertiae" or "force of inertia" and "inertia." For him this was an "internal" rather than an "external" force and so could not – of and by itself – alter a body's state of rest or of motion.

[231] On Darwin and Lyell see Mayr (n. 40 supra); the details of Darwin's transformation are discussed, along with other examples, in my *Newtonian Revolution: With Illustrations of the Transformation of Scientific Ideas* (New York/Cambridge: Cambridge University Press, 1980). This topic is explored also in my forthcoming *Scientific Ideas* (New York: W. W. Norton & Company, 1994).

[232] Mirowski (n. 23 supra, pp. 241–254) has documented the way in which Joseph Bertrand and Hermann Laurent faulted Walras for his mathematical physics, as Laurent and Vito Volterra later faulted Pareto.

[233] *Journal of Economic Literature*, 1991, **29**: 595–596.

[234] But even a severe critic like Hal R. Varian does admit that Mirowski's "thorough search of the writings of Walras, Jevons, Fisher, Pareto, and other neoclassicals . . . has

established, to almost anyone's satisfaction that they recognized that 'utility' had some features in common with the then-current notions of 'energy'."

[235] A. Lawrence Lowell: "An Example from the Evidence of History," in Harvard Tercentenary Conference of Arts and Sciences (1936): *Factors Determining Human Behavior* (Cambridge: Harvard University Press, 1937), pp. 119–132.

[236] Of course, one reason why an analogy may be inappropriate is that it is based on mismatched homology. Another reason might be that the analogy did not advance the subject to the same degree as a rival one.

[237] Henry Guerlac once described it as one of the worst in the English language. Jean T. Desaguliers: *The Newtonian System of the World, the Best Model of Government* (Westminster: A. Campbell, 1728).

[238] John Craig: *Theologiae Christianae Principia Mathematica* (London: impensis Tomothei Child, 1699). A translation of some major extracts by Anne Whitman is published (without the translator's name) as "Craig's Rules of Historical Evidence," *History and Theory: Studies in the Philosophy of History*, Beiheft 4 (The Hague: Mouton, 1964). Craig once suggested to Newton a minor modification of the *Principia*; see I. B. Cohen: "Isaac Newton, the Calculus of Variations, and the Design of Ships," pp. 169–187 of Robert S. Cohen, J.J. Stachel, & M.M. Wartofsky (eds.): *For Dirk Struik: Scientific, Historical, and Political Essays in Honor of Dirk J. Struik* (Dordrecht/Boston: D. Reidel Publishing Company, 1974 – Boston Studies in the Philosophy of Science, vol. 15).

[239] For two centuries and more, Craig's book and its Newton-like laws have usually been presented as an example of the kind of aberration to which Newtonian science may lead. His whole book can, in fact, be considered an extended example of inappropriate analogy. Yet a recent study by Stephen Stigler ("John Craig and the Probability of History: From the Death of Christ to the Birth of Laplace," *Journal of the American Statistical Association*, 1986, **81**: 879–887) has shown that Craig made a serious contribution to applied probability, "that his formula for the probability of testimony was tantamount to a logistic model for the posterior odds."

[240] I have not attempted to rewrite the history of this subject, displayed in many monographs, beginning with Richard Hofstadter: *Social Darwinism in American Thought* (Philadelphia: University of Pennsylvania Press, 1944; rev. ed., Boston: Beacon Press, 1955). Some more recent works are Degler (n. 163 supra) and Robert C. Bannister: *Social Darwinism: Science and Myth in Anglo-American Social Thought* (Philadelphia: Temple University Press, 1979); Howard L. Kaye: *The Social Meaning of Modern Biology: From Social Darwinism to Social Biology* (New Haven: Yale University Press, 1984); Peter J. Bower: *The Eclipse of Darwinism: Anti-Darwinian Evolution Theories in the Decades around 1900* (Baltimore: Johns Hopkins University Press, 1983).

[241] Michael Ruse: "Social Darwinism: Two Roots," *Albion*, 1980, **12**: 23–36.

[242] Spencer: "The Study of Sociology," No. XVI, "Conclusion," *Contemporary Review*, 1873, **22**: 663–677, esp. p. 676.

[243] Stephen Jay Gould: "Shoemaker and Morning Star: A Visit to the Great Reminder reveals some Painful Truths carved in Stone," *Natural History*, December 1990, pp. 14–20, esp. p. 20.

[244] Gould's analogy rests on an imperfect homology. Lamarckian evolution in biology implies not only that each individual may modify his or her inheritance but that such modifications are transmitted to one's offspring. Consider a catastrophe in which all

material culture and all humans over the age of three would be destroyed. In a Lamarckian social world homologous with a Lamarckian biological world, the surviving individuals would have inherited the technological knowledge and skills acquired by centuries of evolutionary development. In the world of nature and of man, however, this would not be the case, as Gould is aware.

[245] Additionally Gould alleges that the Lamarckian mode "of cultural transmission" is responsible for "all the ills of our current environment crisis" as well as "the joys of our confidently growing children."

[246] This example was brought to my attention by Neil Niman at a symposium on Natural Images in Economics. He, however, treats this episode in a wholly different way from mine. See his paper in Mirowski (n. 54 supra).

[247] Armen A. Alchian: "Uncertainty, Evolution and Economic Theory," *Journal of Political Economy*, 1950, **57**: 211–221.

[248] Edith Tilton Penrose: "Biological Analogies in the Theory of the Firm," *The American Economic Review*, 1952, **42**: 804–819, esp. p. 805.

[249] Ibid., p. 807.

[250] Ibid., p. 812.

[251] Armen A. Alchian: "Biological Analogies in the Firm: Comment," *The American Economic Review*, 1953, **43**: 600–603.

[252] Edith T. Penrose: "Rejoinder," ibid., pp. 603–609. Penrose quotes from Alchian's original article to the effect that the "suggested approach embodies the principles of biological evolution and natural selection."

[253] William F. Ogburn: *Social Change with Respect to Culture and Original Nature*, 2d ed. (New York: Viking Press, 1950), Supplement.

[254] Ménard (n. 97 supra; 1988), p. 91.

[255] *Equality of Educational Opportunity*, 2 vols. (Washington, D.C.: Office of Education – U.S. Department of Health, Education, and Welfare – U.S. Government Printing Office, 1966).

[256] Quoted in Frederick Mosteller & Daniel P. Moynihan (eds.): *On Equality of Educational Opportunities: Papers Deriving from the Harvard University Faculty Seminar on the Coleman Report* (New York: Random House, 1972), pp. 4–5.

[257] See the editors' discussion of crude and refined statistics (ibid., pp. 12–14) and also ch. 11 by Christopher S. Jencks on "The Quality of the Data Collected by *The Equality of Educational Opportunity* Survey." The second volume of the Coleman Report consisted of 548 pages of tables of means, standard deviations, and correlation coefficients, as a complement to the 373 pages of the first volume.

[258] In Mosteller & Moynihan (n. 256 supra), p. 33, there is a critique of the statistics and their interpretation. Chapter Four, by James S. Coleman, is on "The Evaluation of Equality of Educational Opportunity."

[259] Mosteller & Moynihan (n. 256 supra), p. 32.

2. THE SCIENTIFIC REVOLUTION AND THE SOCIAL SCIENCES

2.1. THE "NEW SCIENCE" AND THE SCIENCES OF SOCIETY

Ever since the great revolution which produced modern science there has been a hope that a science of society would be created on a par with the sciences of nature. Two early heroes of the Scientific Revolution, Galileo and Harvey, created radical transformations of science – respectively, a physics of motion and a physiology based on the circulation of the blood – which became paradigms for a new social science.[1] Scientific precepts of Bacon and of Descartes were available as guides in this new venture. A primary challenge was to accommodate a new social science to mathematics: either to use classical mathematics for a non-traditional purpose or to introduce a kind of mathematics other than geometry on the Greek pattern. Would-be social scientists could thus find novel ways of dealing with their subject that would transfer to their endeavors the authority of mathematics and the new natural sciences.

In the pre-Newtonian part of the "century of genius" – in the decades that encompass the careers of Galileo, Kepler, Harvey, Bacon, and Descartes – there were a number of earnests of the desired new science of society. Later on in the seventeenth century and during the succeeding century of the "Enlightenment," Newton's spectacular achievement in the *Principia* aroused hopes for a similar science[2] of man and of society, a "human science" of individual behavior and a "social science" of the behavior of large groups. From that day onward, social scientists have been waiting patiently (and sometimes even impatiently) for their "Newton."[3] The history of the social sciences plainly shows that neither the rational mechanics of Newton's *Principia* nor the Newtonian system of the world has ever served successfully as a direct model for engendering a similarly constructed social science.[4] And so, in considering the impact of the natural sciences on the social sciences in the seventeenth century, we shall focus our attention exclusively on the pre-Newtonian decades, taking note of attempts to develop a "science" of government or of the state. We shall examine topics that would later become parts of sociology, political science, economics, or the study of the law. Discussions in all of these areas were to some degree influenced by the revolutionary advances in mathematics and in the physical and biological sciences.

In the early age of the Scientific Revolution, the greatest and most obvious accomplishments were to be seen in the "exact sciences" – mathematics (Descartes, Fermat, and also Galileo), astronomy (Galileo, Kepler), and the physics of motion (Galileo, Descartes, and also Kepler). A comparable revolution in the life sciences was the discovery by Harvey of the circulation of the blood. The mathematical achievements were outstanding because they represented a great conceptual revolution: a new way of thinking based on algebra and analysis rather than the traditional synthetic geometry. The innovations of the new astronomy were both conceptual and observational. Galileo's use of the telescope wholly altered the observational basis of knowledge of the universe, while Kepler introduced non-circular orbits and the concept of sun-planet forces. The most basic alteration in physics occurred in the study of motion, which entailed new conceptual foundations and a mathematicization of nature, to a much greater degree than direct questioning of nature by experiment. From today's point of view the most fundamental change during the early 1600s appears to have been the destruction of the Aristotelian cosmos, the rejection of the traditional concept of the hierarchical nature of space, and the introduction of the new idea of isotropic space, inertial physics, and an infinite – or at least unbounded – universe.[5] The major innovation in the life sciences centered on a radical discovery of the circulation of the blood, based on a conceptual shift made necessary by the introduction of quantitative considerations. Thus the revolutionary changes in science did not consist primarily of the introduction of experiments, as was long believed by historians, but rather was premised on a basic shift of intellectual framework centering on new concepts and the introduction of new mathematical methods.

I have mentioned that Galileo was one of the great heroes of the early Scientific Revolution. In publications, he proclaimed his official position as "philosopher and mathematician." This title accurately recorded the two distinct kinds of science on which his fame was based: empirically based natural philosophy and mathematical science. As empiricist, Galileo was the astronomer who, using the newly invented telescope, showed that the Earth is like the moon and the planets and not unique, thus making the Copernican system philosophically reasonable and hence worthy of serious scientific consideration. His study of the phases of Venus proved that the Ptolemaic system is false.[6] It was as an empiricist that Galileo won renown for experiments of dropping unequal weights from a tower so as to prove that a principal Aristotelian tenet about motion is wrong: bodies do not fall freely in air in the way that Aristotelians supposed.[7] His greatest contribution to physics, however, was not experimental but intellectual: to set forth a new way of thinking about motion, analyzing the problems of natural uniform and accelerated motion in terms of new and clearly defined concepts and principles which he used to develop mathematical laws about speed, distance, and time.[8] In his great book on *Two New Sciences*, Galileo set forth his results concerning motion in a mathematical framework, derived in a geometric style from fundamental definitions and principles. His readers thus saw Galilean physics set in a mathematical structure and did not consider this subject as having been derived from, or even based primarily on, direct experiment.[9] Into the framework of mathematical deduction, Galileo introduced mathematical postulates of physics suited to a new science. Above all, Galileo demonstrated the power of mathematical reasoning applied to abstract or imagined systems that were derived from nature simplified, a method later brought to a high level of fruition by Isaac Newton in his *Principia*.[10] This mathematical method enabled Galileo to transcend the difficulties of the complex physical world of nature by achieving solutions for the ideal case; later, he could introduce some factors of "this" world of "reality." Although Galileo did not essay an application of his science to problems in the social or political arenas, his mathematico-physical model was greatly esteemed by those who strove to produce a mathematically based analysis of social or political affairs.

Along with Galileo, Descartes was generally held in high esteem by social scientists and philosophers during the early years of the Scientific Revolution. Descartes was the author of major works on geometry, optics, and the atmosphere and was a champion of the "mathematical way." He was recognized as a primary founder of the new mathematics, a pioneer in the theory of equations, and an inventor of a new kind of geometry based on algebra, an honor which he shares with Fermat. Descartes was also the author of the celebrated *Discourse on Method* (1637), which was a rival to the precepts of Bacon. Like Bacon, Descartes predicted that the pursuit of natural science would enable human beings to control their environment. His *Principles of Philosophy* embraced the physics of motion (based on the principle of inertia), principles of cosmology, a system of the world, and general philosophy. Additionally, Descartes presented a radical new "science of man," in which all human functions were to be reduced to mechanical actions. This was part of the general "mechanical philosophy," in which nature's operations were to be explained by two "principles": matter and motion.[11]

Early in his career Descartes had a dream which revealed to him the "foundations of the Admirable Science," the way in which he could use the infallible method of mathematics to solve problems of science and philosophy. He envisaged a "universal mathematical science" and even hoped to produce a geometric ethics, a project that he believed might be simpler than constructing a mathematical medicine or physiology.[12] Descartes's human science also drew on his personal experience in making and observing dissections of animals. Furthermore, he devoted a considerable portion of part five of his *Discourse on Method* to a presentation of Harvey's discovery of the circulation of the blood and praised Harvey for his use of observation and experiment.

Harvey's discovery of the circulation of the blood was in keeping with the mathematical spirit of the age, at least to the degree that his great discovery was based on mathematics as well as on a broad range of empirical investigations. Mathematics in the form of quantitative reasoning gave Harvey an early insight into the need for a new physiology and provided a powerful argument for his ideas about the circulation. Harvey's path to discovery, like his presentation in the *De Motu Cordis* of 1628, was solidly based on anatomical investigations (including a great variety of direct observation and experiment), notably in uncovering the function of the valves in the veins and the structure and action of the heart. But readers of *De Motu Cordis* could not help but be impressed by his calculations, which proved that Galen's physiology is inadequate. Harvey found that "the juice of the food that had been eaten" simply would not suffice for the liver to supply "the abundance of the blood that was passed through" the heart. And so, Harvey wrote, "I began to bethink myself" whether the blood "might not have a kind of movement, as it were in a circle." And this, he declared, "I afterwards found to be true."[13]

Harvey's conception of the circulation of the blood was a tremendous advance in human science. He showed that the heart with its valves acts in the manner of a water pump, forcing the blood to flow in a continuous circuit through the animal or human body. This was a direct affront to the prevailing doctrine of Galen, which had dominated medical and biological thought ever since it had been propounded fifteen centuries earlier. Galen had given prominence to the liver as the organ which continually manufactures blood to be sent out through the body and used up as the different parts perform their life functions. But Harvey shifted the physiological primacy of organs from the liver to the heart, whose function, he said, was to a large degree mechanical, forcing blood out through the arteries and drawing blood in from the veins.

Harvey differed from Descartes and Galileo in conceiving that his important scientific discovery could have a direct paradigmatic value in the domain of social affairs. In introducing his great work *De Motu Cordis*, Harvey used his new science of the body to transform the old notion of the body politic. This dramatic example of the use of the new science in a socio-political context occurs at the very beginning of the book, in the long and flowery dedication to the reigning king, Charles I. The following passage expresses Harvey's view unambiguously:

> The heart of creatures is the foundation of life, the prince of all, the sun of their microcosm, on which all vitality depends, from whence all vigor and strength arises. Likewise the King, foundation of his kingdoms and sun of his microcosm, is the heart of the commonwealth, from whence all power arises, all mercy proceeds.

Harvey had no question but that "almost all things human are according to the pattern of man" and "most things in a King are according to that of the heart." Hence "knowledge of his own heart" must be profitable "to a King, as being a divine exemplar of his functions," in accordance with the customary comparison of "great things with small." Since Charles was "placed at the pinnacle of human things" he would be able to "contemplate at one and the same time" both the "principle of man's body" and "the image" of his own "kingly power."[14]

When Harvey wrote of the king's acquisition of knowledge of the heart and its functions, he must have had in mind that Charles had indeed become aware, through Harvey, of this aspect of physiology. Harvey knew Charles personally as a royal physician, and it was through Charles's direct intervention that deer from the royal herd were made available to him for his studies of animal generation. Harvey not only personally instructed the king about the heart and the circulation, as well as about his discoveries in embryology, but he recorded in his *De Generatione Animalium* how he had shown Charles a "punctum saliens" or pulsating point in the uterus of a doe.[15] The king's genuine concern for Harvey's studies of the heart led to the only occasion on which Harvey could actually examine a live human heart beating. After Charles had heard that a son of the Viscount Montgomery had suffered a chest wound that resulted in a permanent open fistula or cavity, permitting direct vision of the interior organs, he instructed Harvey to make a personal examination of the young man. Harvey examined him and was so greatly impressed that he arranged for the young man to be brought to the royal court in order that the king and Harvey might watch the movement of the heart and touch the ventricles while they contracted and expanded, as Harvey himself had already done on his own. Charles was said to have remarked to this young man, "Sir, I wish I could perceive the thoughts of some of my nobilities' hearts as I have seen your heart."[16]

Harvey's comparison of the role of a king and the function of the heart is cast in a traditional mode of thought, the ancient organismic analogy of the "body politic," in which the state was compared to an animal or person, and the sovereign was considered to be the head ruling the body. Some earlier presentations of the body politic used the concept of the heart as a ruler, but others placed the head in this role in the usage still current in our concept of the "head of a state."[17] A few writers on the body politic prior to Harvey had given importance to the heart, but in a framework of Aristotelian or Galenic thought. Thus, in 1565, the surgeon John Halle, who held that "the harte of man [is] a king," declared "the lyver" to be one of "the chief governours under hym," referring to the Galenic principle that the liver is constantly generating new blood from digested food and sending it to the heart.[18] But in Harvey's system the liver was relegated to an inferior position as a result of his own discovery that the blood circulates through the mechanical pumping action of the heart rather than being constantly generated by the liver.[19]

The sovereignty of the heart is a feature of Aristotelian physiology, which even includes the assertion that in the developing embryo the heart is formed before the blood. Harvey's embryological investigations showed, however, that the blood comes into being prior to the embryo heart or other organ, thus revealing the nature of the "punctum saliens" – a feature of the development of the embryo that was to acquire significance in the context of political theory in the writings of James Harrington. Harvey's views on the heart consequently have two features. In his *De Generatione Animalium* the heart is relegated to an inferior position in that it does not appear as the first discernible part in the development of the embryo, but in *De Motu Cordis* the heart acquires a primacy because of its fundamental role in pumping the blood through the animal body. In *Exercitationes Anatomicae de Generatione Animalium* (1651) Harvey made the distinction clear:

> And so being made more sure by those things which I have observed in the egg and in the dissection of living animals, I maintain, contrary to Aristotle, that the blood is the first genital particle, and that the heart is its instrument designed for its circulation. For the function of the heart is the driving on of the blood. . . . [20]

Even when Harvey compared the role of the monarchy to the function of the heart he was not interpreting the heart's primacy in the traditional Aristotelian sense.

Lest it be thought that Harvey introduced the body politic only in the dedication of *De Motu Cordis* and not in the context of his scientific presentation, let me hasten to add that this theme appears again in the text itself, in the concluding chapter seventeen, in which Harvey proves "the hypothesis of the movement and circulation of the blood" by reference to the observable phenomena of the heart and the evidence of "anatomical dissection." The heart is the first organ of the body to appear in a complete form in the developing embryo, Harvey wrote, and it "contains within itself blood, life, sensation and motion before either the brain or the liver was made" or "could perform any function"; to this degree the heart is "like some internal animal." The heart, furthermore, Harvey then declared, is "like the Prince in the Commonwealth in whose person lies the first and supreme power." The heart "governs all things everywhere, and from it as from its origin and foundation in the living creature all power derives and on it does depend."[21]

Harvey's transformation of the traditional organismic analogy of the state (the "body politic") in the context of his own discoveries sanctioned further explorations of political systems based on the new human physiology. Thus the inaugurator of modern physiology introduced his founding treatise with a bold declaration that true science is related to the functioning of the state. I know of no similar statement by any other founder of the new science. Such a sentiment would perhaps come more naturally to Harvey than to a Galileo or a Kepler because the fabric of the human body shows the same kind of complex organization and varied interaction of parts that is found in organized humanity.

2.2. THE SEVENTEENTH-CENTURY GOAL OF A SOCIAL SCIENCE IN MATHEMATICAL FORM (GROTIUS, SPINOZA, VAUBAN)

During the first flowering of the Scientific Revolution in the early seventeenth century, mathematics was the area of most easily discernible achievement. Hence, it is hardly surprising that during the first great century of the Scientific Revolution there were attempts to duplicate the success of these mathematical pioneers by producing a new science of the state or of society in a mathematical mold.

Emulation and application of mathematics took four major forms. The first and perhaps the foremost was the aim of producing works that would display the clarity and certainty of mathematical reasoning, that would be as infallible as Euclidean geometry. The second was the attempt to adopt the actual structural form of presentation: ordered sets of definitions, of axioms and postulates, leading to proved theorems. The third was to apply new mathematical techniques and methods, such as those of algebra and shopkeeper's arithmetic, in order to produce a moral or ethical calculus or a form of social or political mathematics. The fourth was to use numerical social data in the manner that had proved successful in the physical or biological sciences; a corollary was to encourage the collection of such numerical data for this purpose.

The goal of emulating mathematics in creating a new social science may be seen in the thought of Huig (or Huigh) de Groot, or Hugo Grotius (1583–1645), one of the founders of modern international law. Grotius is a particularly significant figure in this context because his reputation was made as a scholarly jurist and his career is not usually associated with the mathematical sciences of the seventeenth century. But in 1636 Grotius corresponded with Galileo in relation to the latter's proposal of a new means of determining longitude at sea, a subject familiar to Grotius, since he had translated (from Dutch into Latin) a work on this topic by the Dutch engineer, Simon Stevin, who was also a friend of his father's.[22] In his letter to Galileo, Grotius expressed his enormous admiration for Galileo's accomplishments, which, he said, "surpass all human endeavor and bring it about that we neither need the writings of the ancients nor fear that any future age will triumph over this one." He would not wish, he continued, "to take to myself the glory of claiming to have been one of your disciples, for it requires great ability to reach that level even when you lead the way." But, he wrote, "if I claim to have been always one of your admirers I will not be speaking falsely." He made his main point in a poetic vein: "I will be happy if in any way I can serve as midwife to your offspring as they come forth into the light of immortality."[23]

Grotius's admiration for a Galilean mathematical physics may be detected in his treatise of 1625, *De Jure Belli ac Pacis*, or *Law of War and Peace*, the work on which his fame was built. In the *Prolegomena* he declared that in writing his treatise he had not considered "any controversies of our own times, either those that have arisen or those which can be foreseen as likely to arise," and he insisted that in this regard he had followed the procedure of mathematicians ("mathematici"). "Just as mathematicians treat their figures as abstracted from bodies," he wrote, "so in treating law I have withdrawn my mind from every particular fact." Grotius evidently believed that his science of international law was as sound and secure as any system of mathematics because he had adopted the same high level of abstraction and accordingly had divorced himself from actual events. He held that his "proofs of things touching the law of nature" were based on "certain fundamental conceptions which are beyond question, "so that no one can deny them without doing violence to himself."[24]

Already in *De Jure Praedae Commentarius*, composed in 1604–1606 although not published in full until 1868, there occurs a statement about "mathematicians" which, despite a slight difference in signification and application, is nevertheless very close to the statement made about "mathematicians" in the famous work published in 1625. In the first chapter of the earlier text, begun when Grotius was only about twenty-one years of age and formulated as a legal brief addressing a particular contemporaneous crisis, the youthful but learned jurist explained his method:

> Just as the mathematicians customarily prefix to any concrete demonstration a preliminary statement of certain broad axioms [*communes quasdam . . . notiones*] on which all persons are easily agreed, in order that there may be some fixed point from which to trace the proof of what follows, so shall we point out certain rules [*regulas*] and laws [*leges*] of the most general nature, presenting them as preliminary assumptions which need to be recalled rather than learned for the first time, with the purpose of laying a foundation upon which our other conclusions may safely rest.

The application of this method in *De Jure Praedae* has been concisely described by Ben Vermeulen:

> The second chapter contains the premisses in the form of nine definitions (*regulae*), in which types of law are described in terms of the gradations of will expressed in a hierarchy of lawgivers, and thirteen precepts (*leges*), which flow from these *regulae*.

Subsequently, various propositions (*conclusiones* and *corollaria*) are derived from the definitions and precepts (chapters III–X). In chapter XI there follows an historical account of the case, which is judged in the light of the *conclusiones* and *corollaria* (chapters XII and XIII . . .).

Thus Grotius is to a certain extent using a method which may be characterized as mathematical or geometrical even if it cannot be regarded as physico-mathematical or arithmetical or quantitative.[25]

Similarly, when Grotius wrote in *De Jure Belli ac Pacis* that he conceived the science of the law of nations in a mathematical mode, he did not intend that law should be given a quantitative base. Rather he meant, as he said, that he would follow a rational procedure: "In my work as a whole I have, above all else, aimed at three things: to make the reasons for my conclusions as evident as possible; to set forth in a definite order the matters which needed to be treated; and to distinguish clearly between things which seemed to be the same and were not." In addition, the Polish scholar Waldemar Voisé has pointed out that in analyzing the concept of justice Grotius adduced "geometrical and arithmetic proportion" and held that for mathematicians "comparative or geometrical" justice "has the name of proportion."[26] Furthermore, conceiving of nature as unalterable, Grotius assumed that neither man nor God could interfere with the necessity of nature's laws. Drawing an example from mathematics, he declared that God himself could not make two times two be anything but four and that God could not alter what had to be in the domain of natural right and natural law. This is akin to a conclusion which Grotius himself recognized as verging on blasphemy, that natural right could exist even if there were no Supreme Being.[27] Grotius thus "freed the concept of natural law from its heteronymous, divine origin" and reduced it to "an element of human nature that can be known by the exercise of reason, in a manner like that which characterizes the rules of mathematics."[28] It may be at least partly because Grotius conceived his system in a mathematical mode and therefore referred to abstractions rather than to real contemporaneous or historical events that he has been criticized as unrealistic by those who have not appreciated the reason for this framework.[29]

The mathematical context of Grotius's work on international law does not receive much attention from today's authorities. Mathematics is not even mentioned in the article on Grotius in the current *International Encyclopedia of the Social Sciences* (1968) or its predecessor, the *Encyclopaedia of the Social Sciences* 1935). In at least one English translation of Grotius's classic work on war and peace, the *Prolegomena* (containing the most explicit discussion of Grotius's mathematical method) are omitted altogether.[30]

To us the characterization of a treatise as mathematical entails either the use of the commonly recognized techniques of mathematical analysis or the introduction of numbers and quantitative data. Accordingly, Grotius's *De Jure Belli ac Pacis* does not appear to us to be mathematical. But in the century of the Scientific Revolution and in the Enlightenment, scholars held the mathematical aspect of Grotius's method to be of the greatest significance. The jurist Christian Thomasius, who published in 1707 a German version of *De Jure Belli ac Pacis*[31] maintained that Hugo Grotius, Thomas Hobbes, and Samuel Pufendorf had distinguished themselves by using the mathematical mode of reasoning about natural law. Thomasius even went so far as to declare that a person who was not a mathematician could never hope to understand the science of natural law. Pufendorf himself explicitly declared that the true science of law had begun only with Grotius and Hobbes, for the reason that they had introduced mathematical reasoning into this subject.[32]

The most celebrated example of the geometrico-mathematical mode of discourse in the age of the Scientific Revolution is the *Ethics* (completed in 1674, but not published until 1677) by Benedict Spinoza (1632–1677), of which the full title reads *Ethica Ordine Geometrico Demonstrata.* Set in a strictly Euclidean framework, this treatise begins with a set of eight numbered definitions and axioms, leading to numbered propositions and their proofs. Later on, there are other sets of numbered definition and axioms, leading to additional propositions and proofs. There are also postulates and lemmas.[33] But although the external form is strictly geometrical or in the style of Euclid (*more geometrico* or *ordine geometrico*), Spinoza does not use the actual techniques of mathematics or geometry in the development of his subject. Nor does his argument in any way depend on numerical data or quantitative considerations.

Spinoza did not employ this geometric form in his other works. But in the *Treatise on Politics* he claimed that he had adopted "the same objectivity as we generally show in mathematical inquiries."[34] That is, in grounding politics on "the real nature of man," he had "taken great care to understand human actions, and not to deride, deplore, or denounce them." In short, he wrote,

I have therefore regarded human passions like love, hate, anger, envy, pride, pity, and the other feelings that agitate the mind, not as vices of human nature, but as properties which belong to it in the same way as heat, cold, storm, thunder and the like belong to the nature of the atmosphere.[35]

Another example of the application of a geometric method to a problem in the social sciences was an essay by Gottfried Wilhelm Leibniz (1646–1716) on the choice of a king of Poland. Titled *Specimen Demonstrationum Politicarum* (specimen or model of political demonstrations), this little work proclaimed through its subtitle that Leibniz had used "a new style of writing intended to produce clear certainty." Published in 1669, eight years before Spinoza's *Ethics*, Leibniz's *Specimen* differs from all similar efforts of that age because his goal was to solve a particular political problem, not to construct an abstract general system.

The *Specimen* is also of interest because it contains a suggestion of a logical calculus of probabilities – in a political context. Although the *Specimen* is not mentioned in many works on Leibniz and is summarily dismissed in others, it did achieve a certain renown in 1921, when John Maynard Keynes began the preface to his treatise on probability by declaring that "the subject matter of this book was first broached in the brain of Leibniz . . . in the dissertation, written in his twenty-third year, on the mode of electing the kings of Poland."[36]

Leibniz develops his subject in a sequence of numbered propositions, interrupted here and there by the introduction of a corollary or lemma. The content of the individual propositions, however, is not generally mathematical. For example, Proposition 9 reads as follows:

Whatever is contrary to LIBERTY is contrary to SECURITY in Poland.
Whatever is contrary to liberty is contrary to the thing most desired by the Poles, *by prop. 3.*
The Poles are a warlike nation, *by prop. 5.*
Whatever is contrary to the desires of a warlike nation is liable to be a cause of war.
Therefore, whatever is contrary to liberty is liable to be a cause of war in Poland.
Therefore, it is liable to be a cause of civil war.
But civil war is dangerous.
Whatever is dangerous is contrary to security.
Therefore, whatever is contrary to liberty is contrary to security in Poland.

By the time the *Specimen* was published the choice had already been made, and the throne was not given to the candidate for whom Leibniz had argued. Thus the *Specimen* is of interest primarily as a pioneering document in the mathematization of political science.

Throughout his life Leibniz was deeply concerned with aspects of political or social science. His goal was to produce a "general science" (*scientia generalis*) that would embrace mathematics and the physical sciences and the social sciences, using a mathematical method for all of these. He aimed also at a "logique civile" or "logique de la vie" in which practical problems, especially legal questions, would be analyzed by a calculus of probabilities. He wanted in particular to provide an easy and certain way to resolve all disputes. "When controversies arise," he wrote, "there will be no more need for disputation between two philosophers than between two accountants." It "will be enough for them to take their pens in their hands and sit down to their sums and say to each other (calling in a friend if they wish): 'Let us calculate.' "[37]

The realm of seventeenth-century mathematical social science embraces not only the works of thinkers whose aim was to emulate the formal structure of geometric systems or to adopt the abstract certainty of mathematical reasoning but also the attempts to produce a numerical base for understanding society and to propose quantitative analyses. In order to have such social numbers it was necessary to have some kind of census.[38] One example will suffice to indicate the growing feeling of need for census numbers: Sébastien Le Prestre de Vauban (1633–1707), Marshal of France under Louis XIV, who has been described as "France's greatest military engineer." Because of his intense concern for, and great use of, statistical or numerical information Vauban has been called "the father of statistics" or "créateur de la statistique."[39] Fontenelle, in the official *éloge* for the Academy of Sciences, said that Vauban was chosen to be an honorary member of the Academy of Sciences as a mathematician because he, more than any other, "had drawn mathematics down from the skies."[40] He wrote a work on a new system of taxes called the *Dixme* or "Tithe."

The desire to have accurate social numbers or census data was part of the seventeenth-century hope of producing a quantitative science of the state and of society. It was a complement to the stated goal of developing a social science that would resemble mathematics both in form and in the certainty resulting from abstraction, from the absence of discussion of issues and events that would arouse human passion. In order to produce a Galilean social science these mathematical aspects were not sufficient. A technique was needed for producing a mathematical interpretation based on numerical social data. Later in the century an attempt was made to use the new mathematical techniques of algebra and shopkeeper's bookkeeping in order to produce such a social science, a "political arithmetick." These antecedent visions of a non-numerical or non-quantitative mathematical social science were important because they heralded the possibility of transferring to the study of society some of the ideals of mathematics which had proved to be fruitful in the new physical sciences.

2.3. POLITICAL ARITHMETIC AND POLITICAL ANATOMY (GRAUNT AND PETTY)

The very notion of introducing mathematics into the social sciences on the model of the natural sciences today suggests much more than the abstract ideas of Grotius or the geometric form of Leibniz's *Specimen* or Spinoza's *Ethics*. Rather, the term "mathematics" at once invokes both the amassing of numerical or quantitative data and the introduction of mathematical techniques: proportions, algebra, graphs, statistical techniques, the calculus, and other types of higher mathematics.

Although various forms of census and of collecting quantitative data on natural resources and other aspects of the economy long antedated the Scientific Revolution,[41] the first useful series of regularly produced social numbers was the London Bills of Mortality, initially issued, on a weekly basis, early in the sixteenth century. They were discontinued, then re-instituted during plague years and, after 1603, were issued more or less regularly, even during years relatively free of any plague or other widespread disease. At first these Bills gave data only on the number of burials. Then christenings were added. Of even more importance for the statistician was the eventual listing by causes of death other than the plague; then, a separation of burials and christenings according to sex.[42]

An important leap forward in mathematical social science occurred when these data were subjected to analysis by John Graunt (1620–1674), a London draper with little formal education, whose reputation was established by the publication in 1662 of a small book entitled *Natural and Political Observations upon the Bills of Mortality*, which secured him election to the Royal Society, the premier scientific organization in Britain.[43] In the dedication, Graunt observes that his work "depends upon the Mathematicks of my Shop-Arithmetick." That is, Graunt did not make use of academic mathematics such as theoretical geometry or abstract number theory. He used business mathematics or accountancy, adding up totals and subtotals, estimating fractions, and analyzing data in the manner of a businessman. He observed, to take one example, that during a period of twenty years, deaths from "*Small Pox, Swine Pox*, and *Measles*, and of *Worms* without *Convulsions*" totalled 12,210, of which he supposed "about 1/2 might be Children under six years old." Some 16,000 of the total of 229,250 deaths were caused by plague. Hence, "about thirty six *per Century* [i.e., percent] of all quick [i.e., live] conceptions died before six years old." Of this total, "*acute* Diseases" other than the plague accounted for "about 50,000, or 2/9 parts." He concluded that this number gave "a measure of the state, and disposition of this *Climate* and *Air* to health."[44]

Graunt made many analyses of his data, according to such factors as years, seasons, and the regions of London. A whole chapter was devoted to "the difference between the numbers of Males and Females." He essayed an estimate of the number of inhabitants of London and tried to determine the rate of population growth, and he compared "causes of death" in "the Country" and in the city. He posed such general questions as: "What proportion die of such general and particular *Casualties*?" "What Years are Fruitful and Mortal, and in what spaces and Intervals they follow each other?" "Why the *Burials* in *London* exceed the *Christenings*, when the contrary is visible in the *Country*?" Above all he urged that "the Art of Governing, and the true *Politicks*," that is, the science of polity, should be based on quantitative data and their analysis. He concluded that such information was needed about the population (including employment), land, and trade. In short, he urged that statecraft be founded on a quantitative base.[45]

Graunt's pioneering analysis soon bore fruit in the "Political Arithmetick" of Sir William Petty, who had strongly influenced Graunt's work. Petty (1623–1687) led an adventurous life, becoming skilled in mathematics and navigation and eventually obtaining a medical degree from Oxford. While serving as army physician in Ireland, he organized a land survey. On his return to England he became a founding Fellow of the Royal Society. He wrote many tracts on economic subjects, of which the most celebrated is the *Political Arithmetick*, published posthumously in 1690.[46] A preliminary statement declares that Petty invented the name "Political Arithmetick" to denote the way in which "the happiness and greatness of the People, are by the Ordinary Rules of Arithmetic, brought into a sort of Demonstration."[47] Petty set forth his method as follows:

The Method I take to do this, is not yet very usual; for instead of using only comparative and superlative Words, and intellectual Arguments, I have taken the course (as a Specimen of the Political Arithmetick I have long aimed at) to express my self in Terms of *Number, Weight*, or *Measure*; to use only Arguments of Sense, and to consider only such Causes, as have visible Foundations in Nature; leaving those that depend upon the mutable Minds, Opinions, Appetites, and Passions of particular Men, to the Consideration of others. . . .[48]

Like Graunt before him, Petty insists on the primacy of numbers and hence arithmetic and its generalization into algebra.[49] This is the new mathematics, not the traditional geometry of academics which goes back to ancient Greece. Further, the topics with which he is concerned (wealth and trade, shipping, taxes, and the cost of maintaining an army) are dealt with in terms of numerical data. In earlier essays in political arithmetic he studied specific questions of housing, hospitals, and populations. For example, finding that the population of London doubles every forty years and the population of "all England" every 360 years, he concluded that the "Growth of London must stop of its self, before the Year 1800" and that "The World will be fully Peopled within the next Two Thousand Years."[50]

Much as we may admire Petty's boldness in setting forth a program for a polity based on social and economic statistics, we must admit that his effort ended in failure. Among the reasons for his lack of success, the primary one was the insufficiency of accurate numerical data. He was, as he admitted, forced to guess the area of a city. He used the reported number of houses and of burials to estimate the population of London, multiplying the number of burials by thirty and the number of houses by six or sometimes by eight, fully aware that in the absence of a proper census he could produce only approximations. Whenever possible he tried to check his estimates by comparisons with other sources – for instance, asserting that his population estimates "do pretty well agree" with such independent data as the poll-tax returns and the bishops' count of communicants[51] – but he usually did not give the actual numbers and in at least one case, as his modern editor observes, "the agreement between Petty's estimate and the bishops' survey is not strikingly close."[52] That he himself was aware of the deficiencies of his numerical results may be seen in a letter to John Aubrey. "I hope," he wrote, "that no man takes what I say about the living and dyeing of men for a mathematical demonstration."[53]

It must also be noted that Petty often used "rash calculations" and even gave "widely varying estimates for the same things." He was also "frequently inaccurate in his use of authorities" and "careless in his calculations"; on "at least one occasion he is open to suspicion of sophisticating his figures."[54] Petty was severely handicapped by not having the technique of graphs and diagrams for the representation of data. The mathematics he would have needed, probability theory, was just then coming into being. Furthermore, he did not really make any fundamental use of algebraic techniques despite his statements to that effect. While, therefore, we may justly laud Petty's vision and the ideal he set forth, we must also admit that the works he produced did not attain the high standard he proclaimed.

An appreciation of Petty's concern for numbers and mathematics must take account of the fact that he was living in an age when the expanding economy of England and the problems of military statecraft were bringing numerical considerations to the fore. As a result of the research of John Brewer and Keith Thomas, we now have a better understanding of the pressure for numerical information by different departments of state in England in Petty's day. These "constituencies" were, as Brewer has shown, "ministers of the crown," who needed information on "all of the various resources of the different departments in order to exercise firm control over government policy"; the Parliament, "both as a policy-maker and as the body dedicated to securing a responsible executive," which needed government statistics; various "occupational groups and special interests directly affected by state policies," which were "eager to learn the grounds on which such decisions were made"; and even the general public, which had developed "a substantial appetite for the sorts of information that only the very considerable resources of the state could provide."[55]

Petty was trained as a physician and recognized the singular importance of anatomy for medicine. He firmly believed that grounding the new science of polity or statecraft on the mathematical analysis of numerical data was an analogue of basing the study of anatomy on dissection, a practise he had learned while a medical student. His most explicit statement of his politico-anatomical method occurs in a posthumously published work on *The Political Anatomy of Ireland* (London, 1691). In the "Author's Preface," Petty asserts that since anatomy is the only sure foundation for knowledge of the "body natural," it follows that an analogous procedure should be used for the "body politick." To "practice" on the body politic without "knowing the *Symmetry, Fabrick,* and *Proportion* of it," would be to act like uneducated healers, "old-women and Empyricks."[56] Furthermore, "*Anatomy* is not only necessary in Physicians," it is a source of valuable knowledge for "every Philosophical person."[57] Petty proudly declared that he had "attempted *the first Essay of Political Anatomy*."[58]

Petty's introduction presents the political problem in purely anatomical terms: proper "*Dissections* cannot be made without variety of proper Instruments." He is, of course, fully aware of the poor quality and even paucity of statistical data on such matters as land holdings, population, rents, wages, and agricultural production. Even so, he concludes, he has been able "to find whereabouts the Liver and Spleen, and Lungs lye," although he has not been able "to discern" in the state "the Lymphatic Vessels, the *Plexus, Choroidus* [or *Choroides*], the *Volvuli* of vessels within the Testicles." Such statements embody Petty's method of analyzing the functions of the state in the manner of an anatomist performing dissections. He was not primarily seeking for analogues of the functions of the state in human physiology, since his primary goal was not to develop a new fashion of the metaphor of the body politic but rather to create a number-based science of the state and to use the tools of mathematics to disclose the laws and principles of statecraft. Reflecting on his endeavors some three centuries later, we may stand in awe at the majesty of his vision and note that, with the exception of economics – no social science has as yet attained the lofty goal of reducing its fundamental laws and principles to an "arithmetic."

2.4. AN INDEPENDENT "CIVIL" SCIENCE BASED ON MOTION (HOBBES)

Whereas Petty attempted to produce a new science of statecraft by combining numerical analysis with a biomedical approach, Thomas Hobbes (1588–1679)[59] aimed to produce a science of politics or of society based on the new science of motion, concepts of mechanics, and the new physiology.[60] Hobbes was magnificently vain about this achievement. He deserved, he wrote "the reputation of having been the first to lay the grounds of two new sciences": one "of *Optiques*, the most curious, and that other of *Natural Justice*, which I have done in my book *De Cive*, the most profitable of all others."[61] We may note, in passing, that in this passage Hobbes was comparing himself (perhaps unconsciously) with Galileo, whose last great work on the science of motion proudly declared that he had created "two new sciences," as expressed in the title, *Discourses and Demonstration Concerning Two New Sciences* (1638). Hobbes's work in optics has not, in fact, gained him a lasting place among the founders of that subject.[62] But his contributions to political science are universally esteemed and have been the source of many centuries of discussion.

Hobbes boasted of his accomplishment as the founder of a new science of human affairs in another declaration, where he began, in a manner reminiscent of Grotius, with a statement of his high regard for Galileo, whose acquaintance he had sought while in Florence (probably in 1635): "Galileus in our time . . . was the first that opened to us the gate of natural philosophy universal, which is the knowledge of the nature of motion."[63] The presentation of a list of physical scientists led him to biology: "Lastly, the science of man's body, the most profitable part of natural science, was first discovered with admirable sagacity by our countryman Doctor Harvey." And now he assessed his own contribution: "Natural philosophy is therefore but young; but civil philosophy is yet much younger, as being no older . . . than my own book *de Cive*."[64]

In creating a new science of politics Hobbes strove to produce a Galilean social science centering on the concept of motion. He was also influenced by Descartes, notably in his use of the Cartesian concept of a "conatus" or "endeavor" to move and in the adoption of a version of the Cartesian notion of inertia.[65] From both Descartes and Galileo, Hobbes derived his strong belief in the certainty of mathematics. The "great masters of the *mathematics*," he wrote "do not so often err as [do] the great professors of the law."[66] Geometry, he declared in *Leviathan*, "is the only science that it hath pleased God to bestow on mankind."[67] In *De Corpore* he sets forth his mathematical principles and applies them in a somewhat Galilean manner to the analysis of various kinds of motion. But he does so on an abstract level more reminiscent of his medieval predecessors[68] and, for example, does not even refer the Galilean law of uniformly accelerated motion to any physical problem of the observed external or physical world of freely falling bodies. It should be added that our faith in Hobbes as a mathematician is weakened, if not destroyed, by his persistent and unwavering belief that he had been able to square the circle.[69]

Hobbes's political goal has been described as an "attempt to create a philosophic system which embraced the science of natural bodies and extended the methods of that science to human actions and political bodies."[70] He was fully convinced that a science of politics or of human society must be similar to a natural science, based on two primary concepts: movement and matter or substance, in accordance with what was known as the "mechanical philosophy." Transferring the importance of motion from the inorganic to the organic world, Hobbes also drew heavily on the discoveries of William Harvey, which must have had a special significance for him insofar as they were based on mathematics, i.e., on quantitative considerations, and centered on the concept of a continual motion. For Hobbes, the circulation of the blood, the "vital motion" (discovered by "Doctor Harvey"), became the very principle of life, "perpetually circulating," so that the "original of life" was said by him to be "in the heart," which he described as being like a great "piece of machinery in which . . . one wheel gives motion to another."[71]

Thus, Hobbes's political system rejects the traditional organismic metaphor in which the state is considered the analogue of an essentially animate being. Learning from Harvey's physiology, reinforced by the Cartesian philosophy, of the degree to which the animal body functions like a complex mechanical device, Hobbes transformed the old concept of the body politic from a purely animate status to that of a great animal machine, acting like an animal but composed of mechanical parts. Drawing directly on Harvey's comparison of the heart to a pump and of the circulatory system to a hydraulic network of pipes or conduits, Hobbes set forth – on the very first page of the introduction to *Leviathan* – the analogy between a machine and an animal or human body. "The *Heart*," he declares, is nothing "but a *Spring*; and the *Nerves*, but so may *Strings*; and the *Joynts*, but so many *Wheeles*, giving motion to the whole Body." He then compares the state or commonwealth, which is "but an Artificiall Man" to a "Naturall" or biological man. In the detail of the comparison he finds that "*Soveraignty* is an Artificiall *Soul*, . . . giving life and motion to the whole body"; the "*Magistrates*, and other *Officers* of Judicature and Execution" are "artificiall *Joynts*"; "*Reward* and *Punishment*" are "the *Nerves*" (by which "every joynt and member is moved to performe his duty"), and so on. Thus, for Hobbes, the purely organic quality of the traditional analogy has become somewhat lost since the body politic has been transformed into a machine that acts and reacts according to physical rather than biological or vital laws and principles.

Hobbes argued that the use of mathematical (i.e., geometrical) reasoning will produce new exact sciences of the mind and of society, i.e., of ethics and politics. In a tripartite claim he held that "*Reason* is the *pace*; Encrease of *Science*, the *way*, and the Benefit of man-kind, the *end*."[72] Hobbes introduced this subject by comparing reasoning and arithmetic. "When a man *Reasoneth*," he wrote, he does nothing else but "conceive a summe totall, from *Addition* of parcels; or conceive a Remainder, from *Substraction* of one summe from another."[73] But reasoning is also analogous to the methods of demonstration that have traditionally "been used onely in Geometry; whose conclusions have thereby been made indisputable."[74] For Hobbes this method consists

> first in apt imposing of Names; and secondly by getting a good and orderly Method in proceeding from the Elements, which are Names, to Assertions made by Connexion of one of them to another; and so to Syllogismes, which are the Connexions of one Assertion to another, till we come to a knowledge of all the Consequences of names appertaining to the subject in hand; and that is it, men call SCIENCE.[75]

It is to be noted that this mode of procedure was said by Hobbes to lead to predictive rules for a human science and so to produce a guide for obtaining predictable results in the domains of ethics or morals and of political action. In short, Hobbes envisioned a social science that would have some of the same qualities of exactness and of predictability as the physical sciences:

> Science is the knowledge of Consequences, and dependance of one fact upon another: by which, out of what we can presently do, we know how to do something else when we will, or the like, another time: Because when we see how any thing comes about, upon what causes, and by what manner; when the like cause come into our power, we see how to make it produce the like effects.[76]

Hobbes firmly believed that if "the moral philosophers had . . . discharged their duty" as "happily" as "the geometricians have very admirable performed their part," then "I know not what could have been added by human industry to the completion of that happiness, which is consistent with human life."[77] For,

> were the nature of human actions as distinctly known, as the nature of quantity in geometrical figures, the strength of avarice and ambition, which is sustained by the erroneous opinions of the vulgar, as touching the nature of right and wrong, would presently faint and languish; and mankind should enjoy such an immortal peace, that . . . there would hardly be left any pretence for war.[78]

Such was the utopian goal of a social or moral science built by the methods of geometry and natural science.

Hobbes's intellectual debt to Galileo and Harvey, and to Descartes, is apparent in his writings and has been the subject of many commentators.[79] His stress on motion and its laws shows that the philosophy of motion espoused by Galileo and by Descartes had made a deep impression on his thought, even to the belief that "the principles of the politics consist in knowledge of the motions of the mind."[80]

Hobbes later on drew up a comparison of the certainty of geometry and of physics and of "civil philosophy." "Geometry," he wrote, "is demonstrable for the lines and figures from which we reason are drawn and described by ourselves," whereas "Civil philosophy is demonstrable because we make the commonwealth ourselves." But, he argued, "because of natural bodies we know not the construction, but seek it from effects, there lies no demonstration of what the causes be we seek for, but only of what they may be."[81] The science of politics, in short, was less certain than geometry but more certain than physics or natural philosophy.

In the opening sentences of *Leviathan*, Hobbes explained that the state is "an Artificiall Animal," and like "all *Automata*" it has "an artificiall life." Thus it is "by Art" that there "is created that great LEVIATHAN called a COMMON-WEALTH, or STATE (in Latine CIVITAS) which is but an Artificiall Man." Then he presents the structure of the state in terms of analogy with the body; for example, corporations are the muscles, public ministers are the organs or nerves, and the problems of the state are the diseases. These analogies were worked up in some detail. One disease "resembleth the Pleurisie" and yet another "infirmity" is much like that caused by "the little Wormes, which Physicisans call *Ascarides*." Another comparison of the irregularities "of a Commonwealth" and the disease "in the Natural Body of man" focusses on a "Distemper" very much like an "Ague," in which "the fleshy parts being congealed, or by venomous matter obstructed; the Veins which by their naturall course empty themselves into the Heart, are not (as they ought to be) supplyed from the Arteries."[82] This is but one of a number of analogies drawn by Hobbes from the Harveyan circulation of the blood and the functioning of the commonwealth. In another, Hobbes said that money is the blood of the commonwealth, observing that the circulation of money is similar to the circulation of "natural Bloud" which "by circulating, nourisheth by the way, every Member of the Body of Man." There are two movements of money, Hobbes observed, one that conveys it "to the Publique Coffers," the other "that Issueth the same out again for publique payments." In this feature "the Artificall Man maintains his resemblance with the Naturall; whose Veins receiving the Bloud from the several Parts of the Body, carry it to the Heart"; there the blood is "made Vitall" and "the Heart by the Arteries sends it out again, to enliven, and enable for motion all the Members of the same."[83]

It must be kept in mind that in Hobbes's presentation, Leviathan or the commonwealth is not supposed to be an animate natural being, but rather "an Artificiall Man" created by the human mind and endowed by the human artificer with functions analogous to those of a natural person. But even though the commonwealth as "Body Politique" is nothing more than a "fictitious" or "artificall" body, its faculties and properties are known through the study of natural physiology (e.g., the work of Harvey) and its actions are known through the study of natural motions (e.g., the work of Galileo and Descartes plus Hobbes's own innovations). The physiology of Harvey had shown that the heart acts in a manner like that of a mechanical pump, thereby providing Hobbes with evidence that the processes of life might be explained mechanically, just as had been taught by Descartes and other advocates of the "mechanical philosophy." Harvey's work thus gave partial sanction to the likening of the functions of animate beings and machines, even though he had never intended that his research world give sanction to the thesis that all bodily functions of animals and human beings were so mechanical that they could be performed by well designed automata.[84]

Hobbes's achievement was to some degree that he used the new discoveries in physiology to transform the organismic concept of the body politic by giving it a mechanical basis in conformity with Descartes's reductionist philosophy. The political and social world of Hobbes is a hybrid kind of organic structure operating mechanically and conceived under the sign of Galileo, Descartes, and Harvey. His system of society was a collection of human beings acting as "mechanical systems of matter in motion" and, like Grotius before him, he broke away "from the traditional reliance on a supposed will or purpose infusing the universe." Tom Sorell suggests that we misinterpret Hobbes if we assume he was "trying to make the scientific status of physics rub off on his civil philosophy," since Hobbes himself suggested that "he regarded civil philosophy as *more* of a science than physics."[85]

2.5. THE NOTION OF A BALANCE: A SOCIAL SCIENCE BASED ON THE NEW PHYSIOLOGY (HARRINGTON)

Hobbes attempted to introduce some aspects of the life sciences into a system of political thought based primarily on the physical science of motion. But James Harrington (1611–1677) took a quite different tack and, in a conscious rejection of Hobbes's methodology, based a socio-political system squarely on the new Harveyan biology, acting as a "scientist of politics".[86] Harrington's work is all the more significant in that he was "the first English thinker to find the cause of political upheaval in antecedent social change."[87] Furthermore, Harrington was ultimately more influential in the sphere of practical politics than Hobbes – or, for that matter, Vauban, Leibniz, Graunt, or Petty – since his doctrines were implemented in the following century, notably in the form of government adopted in the American Constitution.[88]

During the years of the American Revolution and the Constitutional Conventions, many American statesmen were aware that the concept of "balance" in a socio-political context could be traced to James Harrington's *Oceana*. Thus John Adams wrote in his *Defence of the Constitutions* that this political concept was Harrington's discovery and that he was as much entitled to credit for it as Harvey was for the discovery of the circulation of the blood.[89] In this sentiment Adams was echoing the praise given by John Toland, in his edition of Harrington's works, of which there were two copies in Adams's library.[90]

Harrington's principle of the balance was an expression of his radical position that economic forces influence politics, that political power cannot be considered separately from its economic base. Toland put this idea simply and straightforwardly; it is that "*empire follows the balance of property*, whether lodg'd in one, in a few, or in many hands."[91] To use Harrington's own set of examples: if a king owns or controls three quarters of the land in his realm, there is a balance between his monarchical power and his property. But if the king's property was only one quarter, there would be no balance and any absolute monarchical system would be unstable. Similarly, if "the few or a nobility, or a nobility with the clergy," were the landlords, or should "overbalance the people unto the like proportion," the result would be a "Gothic balance," and "the empire" would be a "mixed monarchy." Finally, there is the case in which "the whole people be landlords, or hold the lands so divided among them, that no one man, or number of men, within the compass of the few or aristocracy, overbalance them." In this event, "the empire (without the interposition of force) is a commonwealth."[92]

In Harrington's interpretation, the crisis of the modern world began for England when, under the Tudors, the power of the feudal nobility was broken and the power of land ownership began to be transferred to the people, thus destroying the more or less stable "Gothic balance." He saw the ultimate effect of this change in the English Civil War and held that "the same impersonal forces" were producing political upheavals on the Continent.

Harrington's ideas are set forth primarily in *The Common-Wealth of Oceana*, which was first published in 1656 and has been described as "a constitutional blueprint" and as "little more than a magnified written constitution."[93] In it he proposed a two-body legislature consisting of an elected "Senate" and a body of elected deputies to be known as "the People." He stressed the use of the ballot and even devised an intricate system of indirect elections which contains features that remind us of the American electoral college. He advocated a strict separation of powers and took a strong position on the need for an explicit written constitution. One of his fundamental principles was the rotation of political offices and a strict limit to the time anyone would be allowed to serve. He was primarily concerned with matters of agrarian policy, advocating a strict upper limit on the amount of land anyone could receive by bequest and an even distribution of family lands. Even so brief a catalogue helps us to understand why *Oceana* influenced many of the statesmen who forged the American system of government.

Harrington was a great admirer of William Harvey and declared that his own work was a "political anatomy," which would make it an analogy of the Harveyan anatomy of the animal body.[94] He firmly believed that his dissection of the problems of his age, together with his remedy in proposing new political institutions, constituted more than the traditional sort of historico-political analysis. According to Harrington, it formed an exact equivalent to the physiological anatomy of William Harvey. The "delivery of a model of government," he wrote, must "embrace all those muscles, nerves, arteries and bones, which are necessary unto any function of a well-ordered commonwealth" and is to be likened to "the admirable structure and great variety of the parts of man's body" as revealed by "anatomists."[95] For Harrington this position implied that the political anatomist, like his physiological counterpart, must base his subject on the principles of nature and not merely on one or two examples. William Harvey, he wrote, did not found his discovery of the circulation of the blood on "the anatomy of this or that body" but rather on "the principles of nature."[96]

Harrington's appreciation of the Harveyan physiology was not limited to generalities, but invoked detailed features of the new biological science. He proposed specific anatomical homologies as well as general analogies. In discussing the two chambers of his proposed legislature, Harrington drew directly on Harvey's *De Motu Cordis*, arguing that "the parliament is the heart," which acts like a suction pump, first sucking in and then pumping out "the life blood of Oceana by a perpetual circulation." In this passage we see Harrington's appreciation of Harvey's radical central idea that the heart is a pump. He even followed Harvey in using the mechanistic language of pump technology, and his concept of a continual process of blood circulation is clearly Harveyan. The mere notion of blood flowing in and out does not require more than a superficial acquaintance with the general aspects of the Harveyan circulation. We have seen that Hobbes used such an analogy with respect to money flowing in and out of the national treasury. But Harrington went much deeper into the physiology of the heart and blood. His statement in full is that "the parliament is the heart which, consisting of two ventricles, the one greater and replenished with a grosser store, the other less and full of a purer, sucketh in and gusheth forth the life blood of Oceana by a perpetual circulation."[97] On close analysis, Harrington's analogy has two aspects that draw the attention of the critical reader. The first is the apparent exclusive concentration on the ventricles, to the exclusion of the auricles; the second is the recognition that there is a physically observable difference between the blood ejected from the left and from the right ventricle, as well as that the ventricles are of unequal size.

The critical reader of this paragraph will note that although Harrington fully appreciated that the ventricles suck in and pump (or gush) out blood, he does not mention that the blood which they expel is sucked in from their respective auricles and not directly from the veins. In this context we should note that Harvey explained the circulation as consisting of two partial cycles. In one, the left ventricle pumps blood out of the heart to pass through the aorta into the main system of arteries, returning to the heart through the venous system, and there entering the right auricle; in the other, sometimes known as the "lesser circulation" (or pulmonary circulation or pulmonary transit), the right ventricle pumps out blood through the pulmonary artery and on into the lungs, to return through the pulmonary vein to the left auricle. Thus the heart produces the circulation by means of two auricles and two ventricles, not by two ventricles alone. Hence the historian must raise the question of whether, when Harrington wrote about a two-chambered rather than a four-chambered heart, he was inadvertently showing that his understanding or knowledge of the Harveyan circulation was imperfect or even superficial. This is not an issue of mere pedantry since it has been alleged that he did not really have a deep understanding of science, even of Harvey's work.[98]

In evaluating Harrington's presentation we must keep in mind that in Harvey's day the auricles were usually considered by physiologists and anatomists to be extensions of the veins leading into the heart, continuations of the inferior and superior vena cava. Thus when Harrington concentrated exclusively on the ventricles, the two chambers that expel or pump out blood from the heart, as the principal chambers of the heart, he simply was not concerned with the auricles, the two chambers by which the blood enters the heart after circulating through the arteries and veins. A similar concentration on two chambers of the heart occurs in Descartes's *Discourse on Method* (1637), one of the early works to recognize the validity of Harvey's discovery. Descartes, who had a sound knowledge of the anatomical structure of the heart, recommended that his readers prepare themselves for reading his discussion by witnessing the dissection of "the heart of some large animal" and by having shown to them "the two chambers [*chambres*] or ventricles [*concavitez*] which are there."[99] Harrington was writing in the style of his time when he ignored the auricles and concentrated on the ventricles.

Harrington's invention of the analogy between the heart and the two chambers of a legislature shows both his knowledge of Harveyan science and his originality. Unlike Hobbes, he took cognizance of Harvey's detailed discussion of the physical difference between the blood pumped out by the left ventricle and that pumped out by the right ventricle. Thus his analogy proposes that the two divisions of the legislature have different functions, just as the blood from the two ventricles has different qualities – "the one greater and replenished with a grosser store; the other less and full of a purer."

Harrington's use of *De Motu Cordis* leaves no doubt concerning his conviction that Harvey's discoveries and method had significant implications for the social scientist. I have found, however, that Harrington's knowledge of Harvey went beyond the circulation and the problems of associated physiology and encompassed other aspects of Harvey's science. Those analyses of Harrington's thought that mention Harvey focus exclusively on Harvey's *De Motu Cordia* and the circulation. But the work on which Harvey spent most of his adult years was his *De Generatione Animalium*, published in 1651. It was in this work that Harvey "created" the "term *epigenesis* in its modern-embryological sense," that is, "to denote the formation of the fetus and of animals by addition of one part after another."[100] The spirit of Harvey's research and conclusions was encapsulated in the allegorical frontispiece, showing Zeus opening a pyxis or egg-shaped box from which a variety of animals spring forth. On the egg there is inscribed boldly *Ex ovo omnia*![101] These words do not occur as such in Harvey's text, but they encapsulate his philosophy of generation, that "all things come from an egg." In a day when scientists tended to believe in some variety of preformation,[102] Harvey championed epigenesis and spent most of his research on an attempt to understand mammalian generation in terms of a fertilized ovum.

Not only was Harrington, as a true student of Harvey, acquainted with the ideas of Harvey's *De Generatione*; he also made use of this aspect of Harvey's science in his political thought. In a posthumously published work entitled *A System of Politics*, Harrington wrote: "Those naturalists that have best written of generation do observe that all things proceed from an egg."[103] Here is a direct English translation of the Latin motto in the frontispiece to Harvey's book. In this essay Harrington showed that he had more than just a general notion of epigenesis; he described aspects of the development of the fetus in the egg of a chick in some detail, following the line of Harvey's discoveries, and he used these facts of embryology as the basis of a political analogy.

Harrington began his presentation of embryology with a discussion of the "punctum saliens," or primordial heart of the chick:

> 1. Those naturalists that have best written of generation do observe that all things proceed from an egg, and that there is in every egg a *punctum saliens*, or a part first moved, as the purple speck observed in those of hens; from the working whereof the other organs or fit members are delineated, distinguished and wrought into one organical body.[104]

The "punctum saliens" had been know to embryologists long before Harvey and was considered the starting point of life, the embryonic heart. Aristotle found, as Harvey recorded, that the "punctum saliens" moved.[105]

Aristotle, and later embryologists, believed that the heart was the first organ to be formed in the development of the chick embryo and that the blood was formed later, after the appearance of the liver. It was a feature of the reigning Galenic physiology in Harvey's day that the blood is manufactured by the liver and so could not exist antecedent to the liver. But Harvey demonstrated by careful experiment that in the chick's egg the blood begins its existence before any organ such as the heart or liver takes form. Harvey's studies showed that in the early stages of development of the hen's egg there appears a little reddish purple point "which is yet so exceedingly small that in its diastole it flashes like the smallest spark of fire, and immediately upon its systole it quite escapes the eye and disappears." This red palpitating (or salient) point, the "punctum saliens," was seen to divide into two parts, pulsating in a reciprocating rhythm, so "that while one is contracted, the other appears shining and swollen with blood" and then, when this one "is shortly after contracted, it straightway discharges the blood that was in it" and so on in a continual reciprocating motion.[106] It has been mentioned earlier that Harvey proudly showed the *punctum saliens* to Charles I. Harvey's conclusions have been summed up as follows. The "blood exists before the pulse" and is "the first part of the embryo which may be said to live"; from the blood "the body of the embryo is made," that is, from it "are formed the blood vessels and the heart, and in due time the liver and the brain."[107] Harrington's paragraph number one summarizes Harvey's embryological findings concerning the "punctum saliens" and the way in which the organs develop from it.

Next Harrington introduces a political analogy. His paragraph number two discusses a "nation without government" or one "fallen into privation of form." It is "like an egg unhatched," Harrington wrote, "and the *punctum saliens*, or first mover from the corruption of the former to the generation of the succeeding form, is either a sole legislator or a council."[108] In paragraph number four, Harrington considers the case of "the *punctum saliens*, or first mover in generation of the form" being "a sole legislator," whose procedure – will be "Not only according to nature, but according to art also," beginning "with the delineation of distinct orders of members." This "delineation of distinct organs or members (as to the form of government)," Harrington continues in paragraph number five, is "a division of the territory into fit precincts once stated for all, and a formation of them to their proper offices and functions, according to the nature or truth of the form to be introduced."[109]

In his paragraph number four, Harrington makes a distinction between analysis of the political state by proceeding "according to nature" and "according to art." Again and again in his various writings, he introduced this difference between fundamental science (knowledge of nature) and art (the applications of scientific knowledge). He held that in natural science as well as in political science or the science of society, principles may exist in nature which have not yet been discovered (by science) or applied (in an art). The concept or principle of the political balance, he pointed out, was "as ancient in nature as herself, and yet as new in Art as my writings."[110] Harrington contrasted the originality of his discovery and the eternal nature of the principle he had discovered by comparing his example with that of Harvey. The occasion was a disparagement of *Oceana* by Matthew Wren, who had argued that the principle of the balance was not at all an extraordinary or new discovery made by Harrington but was only a restatement of what had always in some sense be known. Harrington replied by saying that the situation was as if one were to tell "Dr. Harvey that . . . he had given the world cause to complain of a great disappointment in not showing a man to be made of gingerbread, and his veins to run malmesey,"[111] rather than pointing out characteristics of blood known since time immemorial.

Harvey's *De Generatione Animalium* would have been of signal importance for Harrington because of one feature in which it differs markedly from *De Motu Cordis. De Motu Cordis* contains passing remarks on method here and there, but *De Generatione Animalium* presents a general discussion of method – how to do science or how to reason correctly in studying nature – in several short essays that form a preface to the whole work. Here Harvey expressly states that his aim is not merely to make known the new information he has acquired about generation "but also, and this in particular," to "set before studious men a new and, if I mistake not, a surer path to the attainment of knowledge." That way is to study nature and not books, to follow "Nature's lead with their own eyes":

> Nature herself must be our adviser; the path she chalks must be our walk, for thus while we confer with our own eyes and take our rise from meaner things to higher, we shall at length be received into her closet-secrets.[112]

A thinker like Harrington could well have imagined that Harvey was speaking directly to him.

Harvey's methodological essays have two very different aspects which, to a twentieth-century reader, may seem contradictory at first encounter. For not only did Harvey establish certain rules for the direction of research and for producing new knowledge; he also sought to establish a kinship with Aristotle as the master of experimental biological science and first formulator of the mode of biological investigation, the philosopher who had stressed particularly the roles of sensory perception and memory and the path from singulars to universals. It is one of the paradoxes of the history of thought that Harvey should have revolutionized biology while being, to so considerably a degree, an Aristotelian. But even though Harvey constantly praised Aristotle, he did not as a result cease to call attention to Aristotle's mistakes and to correct them.

Again and again in *De Generatione Animalium*, and especially in the methodological essays, Harvey insisted that experience – i.e., direct experiment and observation – is the only way to learn about nature. Thus "sense and experience" are " . . . the source of both . . . Art and Knowledge." Further, he declared, "in every discipline, diligent observation is . . . a prerequisite, and the senses themselves must frequently be consulted." He even went to the extreme of imploring his readers to "take on trust nothing that I say" and calling upon their eyes to be my "witnesses and judges."[113]

The study of nature requires, according to Harvey, not only diligent but repeated observations. For Harvey, observations of nature have two separate aspects. The first is to make a careful description and delineation of each organ or part of the animal or human body; the second to perform what we would call experiments, but which in Harvey's day had not yet been separated out from the more general term "experience" (as is still the case for French and Italian). Harvey's contemporary Kenelm Digby had this latter feature of Harvey's work in mind when he wrote that "Dr. Harvey findeth by experience and teacheth how to make this experience."[114] Stressing the method of induction, Harvey never discussed the great leap of imagination that produces hypotheses to be tested nor did he discuss the directive ideas that are of such great importance in any research program.[115]

Harvey's research program was always guided by his conception of the purpose of anatomical study, displaying one further component that would have been of significance for Harrington. Harvey saw the goal of the anatomist to be a study of the parts of the body in order to determine structures, so that knowledge of such structures could lead to an understanding of functions. By the end of the seventeenth century, this latter aspect of the subject, on which Harvey had left a strong impress, had become generally recognized. That broad compendium of seventeenth-century science, John Harris's *Lexicon Technicum* (1704), defines anatomy as "an Artifical Dissection of an Animal, especially Man," in which "the Parts are severally discovered and explained" in order to serve "for the use of Physick and Natural Philosophy." The new anatomy, in which Harvey was a leading figure, has been characterized as a transformation of the old descriptive or "dead" anatomy into an "anatomia animata," a change from the static "description and drawing up of an inventory of the parts of the human body" to a dynamic understanding of the functions of each part in its structure and of each structure in the processes of life. In short, Harvey proposed what we would call today a study of anatomy in order to produce a physiology.[116]

The political anatomist, following Harvey's precepts, would thus seek detailed information to be organized into political structures – information based on direct experience. William Petty thought that statistical data might replace dissection as a source of experiential information, but Harrington – following the precepts and example of Harvey – held that direct observation was required in the political realm. Harrington's method was like that of the empirical scientist or anatomist, exemplified by Harvey. Experience for the anatomist is the study of actual bodies, living and dead, whereas for the scientist of politics the sources of experience are personal contact (by travel) and reading in the records of history. "No man can be a politician [political scientist]," Harrington wrote, "except he be first an historian or a traveller; for except he can see what must be, or may be, he is no politician." If a man "hath no knowledge in history, he cannot tell what hath been," Harrington pointed out, "and if he hath not been a traveller, he cannot tell what is: but he that neither knoweth what hath been, nor what is, can never tell what must be or what may be."[117]

Harrington's "comparison of the study of politics with anatomy," to quote Charles Blitzer, "was not simply a casual simile," but rather "represented a reasoned belief in the basic likeness of the two disciplines." Both the body politic and the human body are composed of similarly interlocking machine-like structures which function in a co-ordinated manner. Harvey had achieved great success in studying the human body by the method of experience and reason; surely the political anatomist might hope for similar results of significance from the application of a similar method.

In a reply to Matthew Wren, Harrington reminded his critic that "anatomy is an art; but he that demonstrates by this art demonstrates by nature."[118] So is it "in the politics," he wrote, "which are not to be erected upon fancy, but upon the known course of nature," just as anatomy "is not to be contradicted by fancy but by demonstration out of nature." It is "no otherwise in the politics," he concluded, than in anatomy.[119] In short, the study of politics and the study of anatomy were alike because they both sought principles of nature by reason and experience. But Harrington would have agreed with Harvey that one should not be subservient to "the authority of the Ancients." For Harvey the rule was that "the deeds of nature . . . care not for any opinion or any antiquity," that "there is nothing more antient then nature, or of greater authority." Harrington agreed. In the "Epistle Dedicatory," of *De Motu Cordis*, Harvey explained that he did not "profess either to learn or to teach anatomy from books or from the maxims of philosophers" (i.e., from the "works" and "opinions of authors and anatomical writers"), but rather "from dissections and from the fabric of Nature herself."[120] In his second reply to his critics, as represented by Jean Riolan, he referred to confirmation of his ideas by "experiments, observations, and ocular testimony." Harrington, in effect, translated these precepts from human to political anatomy.

Harrington made a deliberate choice in adopting for political science the method of the anatomist, with reliance on direct observation and "experience." That is, he consciously rejected the path of the physical sciences and mathematics. He pilloried Hobbes's use of mathematics in a political context because he particularly abhorred deductive systems that emulated geometry, of which he found a primary example in Hobbes's "ratiocination." Again and again he openly expressed his scorn for what he sometimes called "geometry," sometimes "mathematics," and sometimes "natural philosophy."[121] He made fun of Hobbes for supposing that one could establish a monarchy "by geometry."[122]

Harrington also disdained the physical sciences as a source of models or analogies for politics. He believed that physical science tends to produce abstractions rather than actualities. On this point his views were in harmony with Harvey's. "The knowledge we have of the heavenly bodies," Harvey wrote in the second letter to Jean Riolan, is "uncertain and conjectural." No doubt he had in mind the impossibility of proving whether the Copernican or Ptolemaic or Tychonic system (or any other possible variant) is the true one.[123] In any case, he left no doubt concerning his position: "The example of Astronomie is not here to be followed." The reason, he explained, is that astronomers argue indirectly from observed phenomena to "the causes" and to the reason "why such a thing should be." But for astronomers to proceed as anatomists do in this research, they would have to seek "the cause of the Eclipse" by using direct sense observation, and not by relying on reason alone, and thus would have to be situated beyond the moon.[124]

This astronomical uncertainty can be easily contrasted with the clarity and definiteness of Harvey's proofs. In *De Motu Cordis*, Harvey assembles, one by one, the elements of direct evidence that the heart is pumping blood and that there is a correlation between the systole and diastole of the pulse and the contractions and dilations of the heart, that the blood flows from the heart outward through the arteries and inward toward the heart through the veins, that the valves in the veins permit the passage of blood only in a direction toward the heart. Faced with such anatomical certainty, especially as contrasted with astronomical uncertainty, Harrington made the obvious choice of a scientific model for his political thought.[125]

Harrington's political anatomy resembled Harvey's human and animal anatomy insofar as the purpose was to produce an accurate and careful description and delineation of parts or of anatomical features as a guide to function. Both Harvey and Harrington studied structures with the ultimate goal of producing a general synthesis, in which the working of every structure or organ would become known in relation to its form and structure, so that its actions could be understood as part of the functioning of the body as a living whole. A similar purpose inspirits the anatomical-physiological study of both the animal or human body and the body politic.

Harrington is of special interest in the present context because his political ideas were associated with the science of his day, but especially because his ideas eventually influenced practical policy. His writings and those of Hobbes were among the most widely discussed works of the seventeenth century that embodied the application of science to the socio-political arena. Both *Leviathan* and *Oceana* made extensive use of the new life science. Hobbes combined the organismic analogy with principles of physics and the methods and ideals of mathematics, but Harrington expressly denied that mathematics and mathematical physical science could provide a key to politics. As we have seen, Harrington believed that the principles of politics would be revealed by political anatomy, that is, by following the empirical method of Harvey's anatomy-based life science which yielded principles learned from what Harvey called "the fabric of Nature."[126] Hobbes aimed to transform the centuries-old organismic concept of the body politic into a new metaphor embodying the notion of mechanical system. But Harrington endowed this ancient concept with the qualities of the new Harveyan physiology and so developed a revised biological metaphor and a political science that embodied the chief features of the "Scientific Revolution."

2.6. CONCLUSION

Although the early seventeenth century withnessed a number of attempts to construct one or another social science on principles of mathematics or the natural sciences, no social science of this era ever reached the high level of achievement of Galileo's physics of motion or Harvey's physiology. From the historical perspective these attempts – in the early decades of the Scientific Revolution – to produce a social science that would be equaivalent to the natural sciences are of interest primarily in exhibiting the efforts of many thinkers to understand society and its institutions with the same success that natural scientists achieved with respect to the world of observed nature. In truth, any judgment on the results of these seventeenth-century precursors must take into account that today most social sciences do not exactly resemble their counterpart natural sciences.

Why did these early social scientists seek in the natural sciences a model for their own subjects or attempt to create a social science along the lines of the natural sciences? First of all, there was a natural desire to emulate the works of a Galileo or a Harvey and to share in the accolades given to the natural sciences by using their methods and metpahors in the social domain. Additionally, there was a consensus that the soundest way to create a new science of society was to jettison the traditional reliance on established authorities, such as Plato, Aristotle and the scholastic "doctors," and to start afresh with the new source of authority, nature "herself." Those who practised the natural sciences held that the supreme authority was lodged in nature, and not in the writings of ancient and medieval sages. Seeking an equivalent of the experientially revealed world of nature, social scientists turned to travel or political and social data and to the records of history.

In my analysis I have concentrated on the actual use of the natural sciences by those who aimed to create a social science based upon them. While reading the works of these thinkers, I was impressed again and again by the profundity of their conviction that the natural sciences hold the key to the creation of a science of human behavior and human institutions. Their statements, as I have quoted and summarized them, are strong and unambiguous. Yet it must be admitted that the portions of their works with which I have dealt account for only a small part of the total oeuvre of these writers.[127] In terms of space, and even of direct prominence of presentation, the matters with which I have been concerned may in fact be only a very small part of the treatises published by most of my group of social scientists. Furthermore, it is a fact that today's encyclopedias of the social sciences and general works on the history of political and social theory do not generally mention that Grotius and Harrington, by their own testimony, declared the dependence of their social and political thought on mathematics and the Galilean physics of motion or on the Harveyan physiology. As mentioned, Leibniz's political essay is not cited in general histories of political thought or in even most works on Leibniz. Even Hobbes's use of Galilean physics and Harveyan physiology is discussed only in specialized works. And thus we are lead at once to a pair of fundamental questions. One is whether the use of the natural sciences was really an integral part of their thought as I have alleged, or whether the introduction of the natural sciences was merely a variety of rhetorical flourish so characteristic of that age. The second is why was it that such works as *De Jure Belli ac Pacis* or *Oceana* were not so permeated with the new science that readers could not help but be constantly aware of the foundation in the natural sciences that has been discerned only through detailed historical research?

The second question can be answered more easily and helps us to deal with the first one. If these treatises had been written in such a way that no significant body of the text could be read without some knowledge or understanding of the new mathematics or the natural sciences, then the number of potential readers would have been greatly restricted. Only scholars who were interested in the social sciences and had a scientific or mathematical preparation would have proved equal to the task. Thus, the influence of the works would have been limited. Newton faced a similar situation when he wrote his *Principia*. If he had recast every argument and proof in terms of the algorithm of the calculus which he had invented, the only readers would be those few who had both mastered the new mathematics and were able and willing to adopt a new rational mechanics. On the other hand, if he proceeded in a more geometric and somewhat traditional manner, introducing algebraic formulations of the calculus here and there, he would not frighten away potential readers by facing them with unnecessary chevaux de frise.

The conclusion to which we are led is that the absolute quantity or degree of ubiquity of mathematics or of natural science in the early treatises on the social sciences can not be taken as an index of the degree to which the authors conceived a deep inner dependence on mathematics or on the natural sciences. From today's retrospection what is most significant, therefore, is not the number and extent of the discussions directly involving the natural sciences in the works on political or social science of the seventeenth century, but rather the fact that there are such passages at all. It must have required courage and foresight to attempt to enlarge the domain of the natural sciences by applying the methods of those disciplines to the complex problems of society and of human institutions.

Harvard University

NOTES

[1] Throughout this chapter, I use the term "social science" anachronistically to designate a science of any organized aspect of society. This rubric therefore includes thoughts about society in terms of organization or improvement, international law, statecraft and civil polity, theories of government or the state, and so on. The term "social science" did not come into being until late in the eighteenth century and, as is well known, "sociology" was invented by Comte in the early nineteenth century. See, further, "A Note on 'Social Science' and on 'Natural Science'" in this volume.

[2] In using the word "science" in the discourse of the seventeenth and eighteenth centuries, we must remember that this term did not exclusively designate the area of the natural sciences or mathematics but could be used for any organized branch of knowledge. See §1.1 supra.

[3] Some of the scientists who hoped for a second "Newton" were such diverse specialists as the anatomist and paleontologist Baron Georges Cuvier and the physical chemists Otto Heinrich Warburg, Jacobus Henricus van't Hoff, and Friedrich Wilhelm Ostwald; see

I.B. Cohen: *The Newtonian Revolution* (Cambridge/New York: Cambridge University Press, 1980), p. 294.

[4] See my discussion of "Newton and the Social Sciences: The Case of the Missing Paradigm," to appear in Philip Mirowski (ed.): *Markets Read in Tooth and Claw* (Cambridge/New York: Cambridge University Press, 1993) [in press].

[5] In producing this list I gladly acknowledge the influence of Alexander Koyré. See his *Etudes galiléennes* (Paris: Hermann, 1939); trans. John Mapham as *Galilean Studies* (London: Harvester Press; Atlantic Highlands [N. J.]: Humanities Press, 1978). Also, Koyré's *Metaphysics and Measurement: Essays in Scientific Revolution* (Cambridge: Harvard University Press, 1968). H. Floris Cohen has made a study of the different major interpretations of the causes of the Scientific Revolution; his book on this subject (titled *The Banquet of Truth*) is currently being readied by publication.

[6] The telescope showed that Venus exhibits a sequence of phases which could not occur in the Ptolemaic system. See I.B. Cohen: *The Birth of a New Physics* (revised ed., New York: W.W. Norton & Company, 1985), ch. 4.

[7] The famous experiment of dropping uneven weights from a tower could prove the falsity of the doctrine that heavy bodies fall with speeds proportional to their weights, but could not reveal the laws of motion.

[8] The possible experimental basis of Galileo's discovery of the laws of motion remained secret (that is, confined to Galileo's manuscripts) until our own times, when Stillman Drake began to study the unpublished manuscripts.

[9] Galileo's final presentation of the laws of motion appears in his *Discourses and Demonstrations Concerning Two New Sciences* (1642). In this work, general discussions are in Italian, the language suited for a dialogue in prose, while the mathematical demonstrations are in Latin, and thus set apart from the discussion of general principles.

[10] I have called this method the "Newtonian style," since it was brought to fulfillment and used most effectively by Newton, even thought its roots can be traced back to Galileo. On this subject see the work cited in n. 3 supra and also §1.4 supra.

[11] Although Descartes's contributions to mathematics are presented in every history of the subject, there has never been until recently a full-length and adequate study of Descartes as a physicist. See William R. Shea: *The Magic of Numbers and Motion: The Scientific Career of René Descartes* (Canton [Mass.]: Science History Publications, 1991). On Descartes and the science of motion see René Dugas: *Histoire de la Mécanique* (Neuchâtel: Editions du Griffon, 1950), trans. J.R. Maddox as *A History of Mechanics* (Neuchâtel: Editions du Griffon; New York: Central Book Company, 1955). Also R. Dugas: *La mécanique au XVIIe siécle: des antécédents scolastiques à la pensée classique* (Neuchâtel: Editions du Griffon, 1954), trans. Freda Jacquot as *Mechanics in the Seventeenth Century: From the Scholastic Antecedents to Classical Thought* (Neuchâtel: Editions du Griffon; New York: Central Book Company, 1958).

[12] Descartes expressed this belief in a letter of 15 June 1646 to Pierre Chanut, the French ambassador to Sweden and brother-in-law of Claude Clerselier, the translator of Descartes's works into French and editor of the first collection of Descartes's letters. In this letter, Descartes explained how his "knowledge of physics" has been "a great help to me in establishing sure foundations in moral philosophy." He declared that he had "found it easier to reach satisfactory conclusions on this topic than on many others concerning medicine on which I have spent much more time." Accordingly, "instead of finding

ways to preserve life," he had "found another, much easier and surer way, which is not to fear death." Quoted from Descartes's *Philosophical Letters*, trans. Anthony Kenny (Oxford: Clarendon Press, 1970; Minneapolis: University of Minnesota Press, 1981), p. 196. On Descartes's physiology, see his *Treatise of Man*, trans. Thomas Steele Hall, with introduction and commentary (Cambridge: Harvard University Press, 1972).

[13] William Harvey: *An Anatomical Disputation concerning the Movement of the Heart and Blood in Living Creatures*, trans. Gweneth Whitteridge (Oxford/London: Blackwell Scientific Publications, 1976), p. 75; see also *The Anatomical Exercises of Dr. William Harvey: De Motu Cordis, 1628; De Circulatione Sanguinis, 1649: the First English Text of 1653*, ed. Geoffrey Keynes (London: The Nonesuch Press, 1928), reprinted (without "The Circulation of the Blood") in William Harvey: *Exercitatio Anatomica de Motu Cordis et Sanguinis in Animalibus: Being a Facsimile of the 1628 Francofurti Edition, Together with the Keynes English Translation of 1928* (Birmingham: The Classics of Medicine Library, 1978), p. 58; also "An Anatomical Disquisition on the Motion of the Heart and Blood in Animals," trans. Robert Willis (n. 105 infra), p. 46; also *Movement of the Heart and Blood in Animals: An Anatomical Essay by William Harvey*, trans. Kenneth J. Franklin (Oxford: Blackwell Scientific Publications, 1957), p. 58. These are cited as Whitteridge trans., Willis trans., and Keynes.

On the role of quantitative considerations in the genesis of Harvey's discovery of the circulation, see §2 of the introduction to the Whitteridge translation; also Gweneth Whitteridge: *William Harvey and the Circulation of the Blood* (London: Macdonald; New York: American Elsevier, 1971 – cited as Whitteridge). Also Frederick G. Kilgour: "William Harvey's Use of the Quantitative Method," *Yale Journal of Biology and Medicine*, 1954, **26**: 410–421.

[14] Cf. Keynes 1928 (n. 13 supra), pp. vii–viii; Keynes 1978 (n. 13 supra), pp. v–vi; see also Whitteridge trans. (n. 13 supra), p. 3; Willis trans. (n. 13 supra), pp. 3–4; Franklin trans. (n. 13 supra), p. 3. The Latin text of 1628 is reprinted in facsimile as the first half of Keynes 1978, pp. 3–4. In quoting this passage I use a combined version including some corrections introduced from the original Latin and inclining towards the English translation of 1653, the text which, together with the Latin, would have been available to readers, such as James Harrington, in the seventeenth century.

[15] Whitteridge trans. (n. 105 infra), p. 359; also Willis trans. (n. 105 infra), p. 485. On the significance of the "punctum saliens" in a political context, see §4.5 infra.

[16] Whitteridge (n. 13 supra), pp. 214, 235. Harvey's own description of this episode is given in his *De Generatione Animalium*, Whitteridge trans. (n. 105 infra), pp. 249–251; also Willis trans. (n. 105 infra), pp. 382–384.

[17] On the body politic, see David George Hale: *The Body Politic: A Political Metaphor in Renaissance Literature* (The Hague/Paris: Mouton, 1971), a valuable study even though Hale never considers the relation of the socio-political concept of the body politic to the reigning physiological theories of the body's functioning.

[18] From "The Prologue to the Reader," in John Halle (compiler): *A Very Frutefull and Necessary Briefe Worke of Anatomie, or Dissection of the Body of Man . . . , with a commodious order of notes, leading the chirurgien's hande from all offence and error . . . compiled in three treatises* (London: Thomas Marshe, 1565), published as part of *A Most Excellent and Learned Worke of Chirurgerie, called Chirurgia parva Lanfranchi* . . . (London: Thomas Marshe, 1565).

[19] On Harvey's attitude towards the liver, see Whitteridge (n. 13 supra), esp. p. 142. On the difference between the status assigned to the heart and to the blood by Harvey in *De Generatione* and in *De Motu Cordis*, see n. 21 infra.

[20] Whitteridge trans. (n. 105 infra), p. 242; see also Willis trans. (n. 105 infra) pp. 374–375.

[21] Whitteridge trans. (n. 13 supra), pp. 120, 129–30. In *De Motu Cordis*, Harvey was almost exclusively concerned with the function of the heart as the primary agent producing the circulation and not with the question of whether the heart comes into being in the embryo before the blood. In various other works, and notably in the *De Generatione Animalium*, Harvey made it plain that the blood appears in the development of the embryo before the heart or the liver or any other organ. On Harvey's views concerning the status of the heart and of the blood, especially the difference between *De Generatione* and *De Motu Cordis* and between Harvey's and Aristotle's positions on this topic, see Whitteridge (n. 13 supra), pp. 215–235, and §4.5 infra.

This issue is debated in a set of three articles in *Past and Present*: an original presentation of "William Harvey and the Idea of Monarchy" by Christopher Hill (no. 27, April 1964), a rebuttal by Gweneth Whitteridge (no. 30, April 1965: "William Harvey: A Royalist and No Parliamentarian"), and a reply by Hill (no. 31, July 1965: "William Harvey (No Parliamentarian, No Heretic) and the Idea of Monarchy." These articles are reprinted in Charles Webster: *The Intellectual Revolution of the Seventeenth Century* (London/Boston: Routledge & Kegan Paul, 1974), pp. 160–181, 182–188, 189–196.

Hill's final disclaimer undermines his statement (p. 112) that Harvey's later views have implications which "can only be described as republican – or at best they suggest a monarchy based on popular consent." There is no evidence that Harvey changed his political position from staunch Royalist to supporter of the Commonwealth.

[22] Jacob ter Meulen and P.J.J. Diermanse: *Bibliographie des écrits imprimés de Hugo Grotius* (The Hague: Martinus Nijhoff, 1950), no. 407; Christian Gellinek: *Hugo Grotius* (Boston: Twayne Publishers, 1983), pp. 40, 128 n.78; Hamilton Vreeland: *Hugo Grotius: The Father of the Modern Science of International Law* (New York: Oxford University Press, 1917; reprint, Littleton, Colorado: Fred B. Rothman & Co., 1986), p. 29; M.G.J. Minnaert: "Stevin, Simon," *Dictionary of Scientific Biography*, vol. 13 (New York: Charles Scribner's Sons, 1976), p. 49; Ben Vermeulen: "Simon Stevin and the Geometrical Method in *De Jure Praedae*," *Grotiana*, 1983, **4**: 63–66. Dirk J. Struik: *The Land of Stevin and Huygens: A Sketch of Science and Technology in the Dutch Republic during the Golden Century* (Dordrecht/Boston/London: D. Reidel Publishing Company, 1981), pp. 47, 53, 56.

On Grotius's life and career, see William S.M. Knight: *The Life and Works of Hugo Grotius* (Reading: The Eastern Press, 1925). See also E.H. Kossmann: "Grotius, Hugo," *International Encyclopedia of the Social Sciences*, vol. 6 (New York: The Macmillan Company & The Free Press, 1968); *The World of Hugo Grotius (1583–1645)*: Proceedings of the International Colloquium Organized by the Grotius Committee of the Royal Netherlands Academy of Arts and Sciences, Rotterdam, 6–9 April 1983 (Amsterdam & Maarsen: APA-Holland University Press, 1984); Stephen Buckle: *Natural Law and the Theory of Property: Grotius to Hume* (Oxford: Clarendon Press, 1991); Hedley Bull, Benedict Kingsbury, and Adam Roberts (eds.): *Hugo Grotius and International Relations*

(Oxford: Clarendon Press, 1990); Edward Dumbauld: *The Life and Legal Writings of Hugo Grotius* (Norman: University of Oklahoma Press, 1969); Charles S. Edwards: *Hugo Grotius: The Miracle of Holland: A Study in Political and Legal Thought* (Chicago: Nelson-Hall, 1981).

The Carnegie Endowment for International Peace has published a good translation by Francis W. Kelsey of *De Jure Belli ac Pacis Libri Tres* (Oxford: Clarendon Press; London: Humphrey Milford, 1925 – The Classics of International Law, no. 3, vol. 2); in the same series (no. 3, vol. 1) is a facsimile reproduction of the Latin edition of 1646 (Washington: Carnegie Institution of Washington, 1913). See also Hugo Grotius: *De Jure Belli ac Pacis Libri Tres*, ed. and trans. William Whewell, 3 vols. (Cambridge: John W. Parker, London, 1853). In this edition, the English translation (an abridged version) appears at the bottom of the page underneath the Latin text.

[23] Galileo Galilei: *Le Opere*, vol. 16 (Florence: Tipografia Barbèra, 1905 and later reprints), pp. 488–489, a letter from Hugo Grotius in Paris to Galileo, written in September 1636; also in Hugo Grotius: *Briefwisseling*, vol. 7, ed. B.L. Meulenbroek (The Hague: Martinus Nijhoff, 1969 – Rijks Geschiedkundige Publicatiën, Grote Series, 130), pp. 398–399.

Grotius wanted to find an asylum for Galileo when the latter had been condemned by the Inquisition. See Hugo Grotius: *Briefwisseling*, vol. 5, ed. B.L. Meulenbroek, (The Hague: Martinus Nijhoff, 1966 – Rijks Geschiedkundige Publicatien Grote Serie 119), pp. 489–490. See also Giorgio de Santillana: *The Crime of Galileo* (Chicago/London: The University of Chicago Press, 1955; Midway reprint, 1976), p. 214 n. 17.

[24] Kelsey trans. (n. 22 supra), pp. 23, 29–30; also Whewell trans. (n. 22 supra), vol. 1, pp. lxv, lxxvii.

[25] Hugo Grotius: *De Jure Praedae Commentarius: Commentary on the Law of Prize and Booty*, vol. 1: A Translation of the Original Manuscript of 1604 by Gladys L. Williams with the collaboration of Walter H. Zeydel (Oxford: at the Clarendon Press; London: Geoffrey Cumberlege, 1950 – Publications of the Carnegie Endowment for International Peace, Washington; The Classics of International Law, no. 2, vol. 1; also reprinted, New York: Oceana Publications; London: Wiley & Sons, 1964), p. 7; Hugo Grotius: *De Jure Praedae Commentarius*, vol. 2: The Collotype Reproduction of the Original Manuscript of 1604 in the Handwriting of Grotius (Oxford: at the Clarendon Press; London: Geoffrey Cumberlege, 1950 – Publications of the Carnegie Endowment for International Peace, Washington; The Classics of International Law, no. 2, vol. 2), f. 5^r; Ben Vermeulen (n. 22 supra), p. 63 (with specific mention of his not discussing "the non-juridical chapters XIV and XV); cf. also Alfred Dufour: "L'influence de la méthodologie des sciences physiques et mathématiques sur les fondateurs de l'Ecole du Droit naturel moderne (Grotius, Hobbes, Pufendorf)," *Grotiana*, 1980, **1**: 33–52, esp. 40–44; Alfred Dufour: "Grotius e le droit naturel du dix-septième siècle," in *The World of Hugo Grotius* (n. 22 supra), pp. 15–41, esp. 22–23; Peter Haggenmocher: "Grotius and Gentili: A Reassessment of Thomas E. Holland's Inaugural Lecture," in Bull (n. 22 supra), pp. 142–144, 162; C.G. Roelofsen, "Grotius and the International Politics of the Seventeenth Century," in Bull, pp. 99, 103–111. It must also be said that the mathematical aspect should not be overemphasized; Knight (n. 22 supra), for example, thinks of the procedure in *De Jure Praede* as scholastic (p. 84). The revised twelfth chapter of *De Jure Praedae* was published in 1609 as *Mare Liberum*. The manuscript of *De Jure Praedae* was dis-

covered in 1864 and finally published in full as *De Jure Praedae Commentarius*, ed. H.G. Hamaker (The Hague: Martinus Nijhoff, 1868). See Meulen and Diermause (n. 22 supra), nos. 541, 684. It should be noted that the geometrical form of *De Jure Praedae* is much less striking than that of Leibniz in his *Specimen* (n. 36 infra). The two documents are comparable, however, because of their invocation and use of mathematical method, their addressing of a specific contemporary crisis, and the youth of their authors.

[26] Kelsey trans. (n. 22 supra), p. 29; also Whewell trans. (n. 22 supra), vol. 1, p. lxxvii. Voisé (n. 30 infra), p. 86.

[27] Kelsey trans. (n. 22 supra), pp. 40, 13; Whewell trans. (n. 22 supra), vol. 1, pp. 12, xliv–xlvi. See also Ernst Cassirer: *The Myth of the State* (New Haven: Yale University Press, 1946), p. 172; reprint (Garden City, N.Y.: Doubleday & Company [Doubleday Anchor Books], 1955), p. 216; also, e.g., Hendrik van Eikema Hommes: "Grotius on Natural and International Law," *Netherlands International Law Review*, 1983, **30**: 61–71, esp. 67.

[28] Voisé (n. 30 infra), p. 86. Cf. Jerzy Lande, *Studia z filozofii prawa*, ed. Kazimierz Opalek & Jerzy Wróblewski (Warsaw: Panstwowe Wydawnictwo Naukowe, 1959), pp. 537–543.

[29] Johan Huizinga: *Men and Ideas: History, the Middle Ages, the Renaissance,* trans. James S. Holmes and Hans van Marle (New York: Meridian Books, 1959), pp. 332–333, 337–338; and Voisé (n. 30 infra), p. 85.

[30] Hugo Grotius: *The Rights of War and Peace*, trans. A.C. Campbell (Washington/London: M. Walter Dunne, 1901; reprint, Westport, Conn.: Hyperion Press, 1979).

Cassirer (n. 27 supra, p. 165), of course, was aware of Grotius's admiration for Galileo and Grotius's reliance on the method of mathematics, but even he did not deal in full with these topics. The only work which I have encountered which seriously addresses this aspect of Grotius's career is Waldemar Voisé: *La réflexion présociologique d'Erasme à Montesquieu* (Wroclaw: Zaklad Narodowy Imienia Ossolinskich, Wydawnictwo Polskiej Akademii Nauk, 1977), esp. pp. 84–87. But even Voisé does not explore fully the consequences of Grotius's choice of a mathematical model.

[31] Voisé (n. 30 supra), p. 88.

[32] Ibid., pp. 88–89.

[33] Spinoza's *Ethics*, published posthumously, is available in a number of different English editions. A good, recent reference work on Spinoza's *Ethics* is Jonathan Bennett: *A Study of Spinoza's Ethics* (Indianapolis: Hackett Publishing, 1984). Spinoza's work on Descartes's *Principles of Philosophy* was translated by Halbert Hains Britan (Chicago: The Open Court, 1905).

[34] Benedict Spinoza: *The Political Works*, ed. and trans. A.G. Wernham (Oxford: Clarendon Press, 1958), p. 263. This volume contains a very valuable historical and critical study plus the complete text of the *Tractatus Politicus* and a translation of the major portions of the *Tractatus Theologico-Politicus*.

[35] Idem.

[36] See John Maynard Keynes: *A Treatise on Probability* (London: Macmillan and Co., 1921; reprint, New York: AMS Press, 1979), p.v.; also reprinted as vol. 8 of *The Collected Writings of John Maynard Keynes* (London: Macmillan for the Royal Economic Society, 1973), p. xxv. Leibniz's *Specimen Demonstrationum Politicarum pro Eligendo Rege*

Polonorum novo scribendi genere ad claram certitudinem exactum is published in the original Latin in *Sämtliche Schriften und Briefe*, series 4, vol. 1, ed. Prussian Academy of Sciences (Darmstadt: Otto Reichl Verlag, 1931,), pp. 3–98; for editorial comment, see this volume, pp. xvii–xx, and vol. 2, ed. German Academy of Sciences at Berlin (Berlin: Academie-Verlag, 1963), pp. 627–635. This text is not included in Patrick Riley (ed.): *Political Writings of Leibniz* (Cambridge/London/New York: Cambridge University Press, 1972), nor is there a reference to it in the editor's introduction and notes.

An exception to the general rule is Eric Aiton: *Leibniz: A Biography* (Bristol: Adam Hilger, 1985), which has a brief discussion of the *Specimen*; more typical of those works on Leibniz that mention the *Specimen* at all is C.D. Broad: *Leibniz: An Introduction*, ed. C. Lewy (Cambridge: Cambridge University Press, 1975), p. 3: "Among his minor achievements was to produce a geometrical argument to prove that the electors to the monarchy of Poland ought to choose Philip Augustus of Neuburg as king."

I have completed a full-length study of Leibniz's *Specimen* and its significance, to be published (in 1992) in *History and Philosophy of Science.*

[37] Godfried Wilhelm Leibniz: *Die Philosophischen Schriften*, ed. C.I. Gerhardt, vol. 7 (Berlin: Weidmannsche Buchandlung, 1890), p. 200 (trans. mine). The strength of Leibniz's conviction is revealed by the number of versions which he made of this passage: cf., e.g., ibid., pp. 26, 64–65, 125; Eduard Bodemann: *Die Leibniz-Handschriften der Königlichen Öffentlichen Bibliothek zu Hannover* (Hanover: Hahn, 1895 [not 1889]; reprint, Hildesheim: Georg Olms Verlagsbuchhandlung, 1966), p. 82; Leibniz: *Opera Omnia*, ed. Ludovicus Dutens, vol. 6, part 1 (Geneva: Apud Fratres De Tournes, 1768; also reprint, Hildesheim/Zurich/New York: Georg Olms Verlag, 1989), p. 22; Leibniz: *Opuscules et fragments inédits de Leibniz: extraits des manuscrits de la Bibliothèque royale de Hanovre*, ed. Louis Couturat (Paris: Félix Alcan, Éditeur, 1903), pp. 155–156, 176. See also Louis Couturat: *La logique de Leibniz d'après des documents inédits* (Paris: Félix Alcan, Éditeur, 1901), p. 141.

[38] Hyman Alterman: *Counting People: The Census in History* (New York: Harcourt, Brace & World, 1969), esp. pp. 45–47.

[39] Henry Guerlac: "Vauban," *Dictionary of Scientific Biography*, vol. 13 (New York: Charles Scribner's Sons, 1976), p. 590, 591; on Vauban's work on "statistique et prévision," see Michel Larent: *Vauban: un encyclopédiste avant la lettre* (Paris: Berger-Levrault, 1982), pp. 132–160. Vauban's *Dixme royale*, originally published in 1707, is available in a scholarly edition, based on the original printing plus various manuscripts, E. Coornaert (ed.): *Projet d'une dixme royale, suivi de deux écrits financiers* (Paris: Librairie Félix Alcan, 1933).

[40] Francisque Bouiller (ed.): *Eloges de Fontenelle* (Paris: Garnier Frères, 1883), p. 28.

[41] See Alterman (n. 38, supra); also Helen M. Walker: *Studies in the History of the Statistical Method: With Special Reference to Certain Educational Problems* (Baltimore: Williams & Wilkins, 1929; reprint, New York: Arno Press, 1975), p. 32. This valuable work should be supplemented by Stephen M. Stigler: *The History of Statistics: The Measurement of Uncertainty before 1900* (Cambridge/London: The Belknap Press of Harvard University Press, 1986); and John A. Koren: *The History of Statistics: Their Development and Progress in Many Countries* (New York: The Macmillan Company, 1918; reprint, New York: Burt Franklin, 1970).

For an understanding of the numeracy of the age of Graunt and Petty, see especially

John Brewer: *The Sinews of Power: War, Money and the English State, 1688–1783* (New York: Alfred A. Knopf, 1989; paperback reprint, Cambridge: Harvard University Press, 1990), ch. 8, "The Politics of Information: Public Knowledge and Private Interest." The Knopf edition is used here; there are also two British editions: London: Century Hutchinson, 1988; London/Boston: Unwin Hyman, 1989. See, further, Keith Thomas: "Numeracy in Early Modern England," *Transactions of the Royal Historical Society*, 1987, **37**: 103–132.

[42] A thorough account of the Bills of Mortality may be found in Charles Henry Hull (ed.): *The Economic Writings of Sir William Petty, together with Observations upon the Bills of Mortality more probably by Captain John Graunt*, 2 vols. continuously paginated (Cambridge: Cambridge University Press, 1899; reprint, Fairfield [N.J.]: Augustus M. Kelley, 1986), pp. lxxx–xci.

[43] The fifth edition (London, 1676) of Graunt's *Observations* is reprinted in Hull's edition of Petty, pp. 314–435. The first edition (London, 1662) has been reprinted in facsimile (New York: Arno Press, 1975). Hull (pp. xxxiv–xxxviii) has assembled all the information about Graunt's life and on the authorship of the *Observations upon the Bills of Mortality*. Hull concludes that Graunt was "in every proper sense the author of the *Observations*," but he assembles evidence that Petty had an important role in the actual composition of the book, in addition to providing Graunt with medical and other information.

A later analysis of this question by Major Greenwood: *Medical Statistics from Graunt to Farr* (Cambridge: Cambridge University Press, 1948; reprint, New York: Arno Press, 1977), contains (pp. 36–39) an updated discussion of whether Graunt wrote "the book published over his name." Greenwood reviews the history of the question and lists in chronological order some studies relating to this controversy from 1925 to 1937. He concludes that Graunt was indeed the author but that a life-table in Graunt's *Observations* may have originated with Petty, the argument being that it is "far too conjectural to have been the work of so cautious a reasoner as Graunt."

[44] The importance of climate and air for health was a major feature of medical thought from the time of Hippocrates, whose treatise on "Airs, Waters, Places" continued to exert a significant influence up to the end of the eighteenth century.

[45] Hull (n. 42 supra) discusses "Graunt and the Science of Statistics" on pp. lxxxv–lxxix. Stigler (n. 41 supra), p. 4, remarks that Graunt's *Observations* "contained many wise inferences based on his data, but its primary contemporary influence was more in its demonstration of the value of data gathering than on the development of modes of analysis."

[46] Petty's *Political Arithmetick* is reprinted in volume one of Hull's edition (n. 42 supra). An important recent study of Petty is Peter H. Buck: "People Who Counted: Political Arithmetic in the Eighteenth Century," *Isis*, 1982, **73**: 28–45. Petty's work is also discussed in histories of probability and statistics, e.g., Walker (n. 41 supra).

Petty is esteemed today for his writings on economics as much as for his work on demography and political arithmetic. In economics, Petty is noted for an early statement of the doctrine of "division of labor." See William Letwin: *The Origins of Scientific Economics: English Economic Thought, 1660–1776* (London: Methuen & Co., 1963; reprint, Westport: Greenwood Press, 1963), ch. 6.

An extremely valuable resource for Petty studies, containing a wealth of information drawn from otherwise unused manuscript sources, is Lindsay Gerard Sharp: *Sir William Petty and Some Aspects of Seventeenth Century Natural Philosophy* (Unpublished D. Phil. Thesis, Faculty of History, Oxford University, deposited in the Bodleian Library 2.2.77). Scholars in many fields will regret that this important study was never published.

A useful reference source is Sir Geoffrey Keynes: *A Bibliography of Sir William Petty, F.R.S., and of Observations on the Bills of Mortality by John Graunt, F.R.S.* (Oxford: Clarendon Press, 1971).

[47] Quoted from Lord Edmund Fitzmaurice: *Life of Sir William Petty, chiefly from Private Documents hitherto unpublished* (London: John Murray, 1895), p. 158. Petty used the term "political arithmetick" even earlier, in print, in his *Discourse of Duplicate Proportion* (London, 1674), and, at an earlier date, in a letter to Lord Anglesey, 17 December 1672. See Hull (n. 42 supra), p. 240n.

[48] *Political Arithmetick*, preface, in Hull (n. 42 supra), p. 244.

[49] In a letter to Edward Southwell, 3 November 1687, Petty described at length what algebra is. After giving an explanation of the principles and a number of examples, he concluded with a brief history, tracing the origins to Archimedes and Diophantus but noting that "Vieta, DesCartes, Roberval, Harriot, Pell, Outread, van Schoten and Dr. Wallis have done much in this last age." He then noted that algebra "came out of Arabia by the Moores into Spaine and from thence hither, and W[illiam] P[etty] hath applyed it to other then purely mathematicall matters, viz: to policy by the name of *Politicall Arithmitick*, by reducing many termes of matter to termes of number, weight, and measure, in order to be handled Mathematically." These two remarks of Petty are excerpted from the Petty-Southwell Correspondence in *The Petty Papers: Some Unpublished Writings of Sir William Petty*, ed. by Marquis of Lansdowne, 2 vols. (London: Constable & Company; Boston/New York: Houghton Mifflin Company, 1927), vol. 2, pp. 10–15; cf. pp. 3–4.

[50] Hull (n. 42 supra), p. 460.

[51] Ibid.

[52] Ibid., p. lxvii, n. 6.

[53] Ibid., p. lxviii.

[54] Ibid.

[55] Brewer (n. 41 supra), p. 223.

[56] Hull (n. 42 supra), pp. 451–478, esp. p. 473.

[57] Ibid., p. 501.

[58] Ibid., pp. 521–544.

[59] Of all the thinkers presented in this chapter, Hobbes is the one most familiar to students of social or political thought. Furthermore, it is generally known that Hobbes based his system on the new physics of motion, but less attention has been paid to his use of Harveyan physiology. Hence my presentation of Hobbes's use of the natural sciences stresses the biomedical basis of his political thought rather than his use of mathematics and the physical sciences.

There are many good presentations of the thought of Hobbes, among them Leo Strauss: *The Political Philosophy of Hobbes: Its Birth and Its Genesis*, trans. from the German manuscript by Elsa M. Sinclair (Oxford: The Clarendon Press, 1936; Chicago: University of Chicago Press, 1966); Arnold A. Rogow: *Thomas Hobbes: A Radical in the Service of Reaction* (London/New York: W.W. Norton & Company, 1986). There is much to be learned from two volumes by C.B. Macpherson: *The Political Theory of Possessive Individualism, Hobbes to Locke* (Oxford: Clarendon Press, 1962), and *Democratic Theory: Essays in Retrieval* (Oxford: Clarendon Press, 1973).

Especially important in the present context is an essay by J.W.N. Watkins: "Philosophy and Politics in Hobbes," *Philosophical Quarterly*, 1955, **5**: 125–146; expanded into the book *Hobbes's System of Ideas: A Study in the Political Significance of Philosophical Theories* (London: Hutchinson & Co., 1965; 2d ed., 1973). Also Thomas A. Spragens: *The Politics of Motion: The World of Thomas Hobbes* (London: Croon Helm, 1973); and M.M. Goldsmith: *Hobbes's Science of Politics* (London/New York: Columbia University Press, 1966).

Also David Johnston: *The Rhetoric of Leviathan: Thomas Hobbes and the Politics of Cultural Transformations* (Princeton: Princeton University Press, 1986); Tom Sorell: *Hobbes* (London/New York: Routledge & Kegan Paul, 1986 – The Arguments of the Philosophers); Richard Tuck: *Hobbes* (Oxford/New York: Oxford University Press, 1989 – Past Masters); and Frithiof Brandt: *Thomas Hobbes' Mechanical Conception of Nature* (Copenhagen: Levin & Munksgaard, 1928).

[60] Hobbes's *Leviathan*, his major work, is available in many editions and reprints, among them the Pelican Classics edition, ed. C.B. Macpherson (Harmondsworth: Penguin Books, 1968). The most recent edition, ed. Richard Tuck (Cambridge/New York: Cambridge University Press, 1991) has indexes of subjects and of names and places and a concordance with earlier editions.

The writings of Hobbes have been collected in two sets – Sir William Molesworth (ed.): *The English Works of Thomas Hobbes*, 11 vols. (London: John Bohn, 1839–1845; reprint, Aalen [Germany]: Scientia, 1962); Sir William Molesworth (ed.): *Thomae Hobbes Malmesburiensis Opera Philosophica Quae Latine Scripsit Omnia*, 5 vols. (London: John Bohn, 1839–1845; reprint, Aalen [Germany]: Scientia, 1961). There are also articles on Hobbes in the *Encyclopaedia of the Social Sciences*, vol. 4 (New York: The Macmillan Company, 1937), and the *International Encyclopedia of the Social Sciences*, vol. 6 (U.S.A.: The Macmillan Company & The Free Press, 1968).

[61] *English Works* (n. 60 supra), vol. 7, pp. 470–471.

[62] On Hobbes's optics, se Alan E. Shapiro: "Kinematic Optics: A Study of Wave Theory of Light in the Seventeenth Century," *Archive for History of Exact Sciences*, 1973, **11**: 134–266.

[63] "Epistle Dedicatory," *De Corpore*, in *English Works* (n. 60 supra), vol. 1, p. viii.

[64] Ibid. It should be noted that in these two referrences to his own place in history, Hobbes refers specifically to his *De Cive*, not to *Leviathan*.

[65] On the Cartesian notion of inertia, see A. Koyré: *Galilean Studies* (n. 5 supra), part 3, "Descartes and the Law of Inertia." See also the works by R. Dugas and W. Shea cited in n. 11 supra.

[66] *English Works* (n. 60 supra), vol. 6, p. 3.

[67] *Leviathan*; ch. 4, Tuck ed. (n. 60 supra), p. 28. Hobbes learned geometry only late in life and was never a real master of the subject.

[68] On the style of the writers on mechanics of the late Middle Ages, see Marshall Clagett: *The Science of Mechanics in the Middle Ages* (Madison: University of Wisconsin Press, 1959); also John E. Murdoch and Edith D. Sylla: "The Science of Motion," pp. 206–264 of David C. Lindberg (ed.): *Science in the Middle Ages* (Chicago/London: University of Chicago Press, 1978).

[69] The mathematician John Wallis kept up a continual exposure of Hobbes's attempts to square the circle. Although it had not then been proved that the squaring of the circle was impossible, no mathematician "worthy of his salt" in the seventeenth century would believe such a feat to be possible. On Wallis's attack on Hobbes for his attempts to square the circle, see J.F. Scott: *The Mathematical Work of John Wallis* (London: Taylor and Francis, 1938), pp. 166–172.

[70] Goldsmith (n. 59 supra), p. 228.

[71] *English Works* (n. 60 supra), vol. 1, pp. 406–407; see *Leviathan*, Tuck ed. (n. 60 supra), p. 3; Spragens (n. 59 supra), p. 69.

[72] *Leviathan*, ch. 5; Tuck ed. (n. 60 supra), p. 36.

[73] Ibid., p. 31.

[74] Ibid., p. 34.

[75] Ibid., p. 35.

[76] *English Works* (n. 60 supra), vol. 3, p. 35.

[77] Ibid., vol. 2, p. iv.

[78] Ibid.

[79] See Brandt, Goldsmith, Sorell, Tuck, Watkins (see n. 59 supra).

[80] Spragens (n. 59 supra), *De Corpe*, I. vi, 7, English Works, vol. 1, p. 74.

[81] *Six Lessons to the Professors of Mathematics* (Ep. Ded.), *English Works* (n. 60 supra), vol. 7, p. 184.

[82] *Leviathan*, ch. 29; Tuck ed. (n. 60 supra), pp. 228–230.

[83] Ibid., ch. 24; Tuck ed. (n. 60 supra), pp. 174–175. Although Hobbes does say that the blood that passes through the heart, before being pumped out again into the arteries, "is made Vitall," he does not indicate that the blood entering the heart from the parts of the body is made to pass out into the lungs and then back again into the heart before going out into the parts of the body once again. He does not make use of Harvey's observation that the alteration of the blood does not occur as it passes through the heart but is a result of the pulmonary transit or passage through the lungs in what is sometimes known as the lesser circulation or pulmonary circulation. Nor does Hobbes indicate that there is an observable physical difference between the blood entering the heart from the lungs and the blood coming into the heart from the various other parts of the body.

[84] Leonora Cohen Rosenfield: *From Beast-Machine to Man-Machine: Animal Soul in French Letters from Descartes to La Mettrie* (New York: Oxford University Press, 1941).

[85] Tom Sorell: "The Science in Hobbes's Politics," pp. 67–80 of G.A.J. Rogers & Alan Ryan (eds.): *Perspectives on Thomas Hobbes* (Oxford: Clarendon Press, 1988), esp. p. 71.

[86] C.B. Macpherson: "Harrington's 'Opportunity State,'" reprinted from *Past and Present* (no. 17, April 1960) in Webster (n. 19 supra), pp. 23–53, esp. p. 23. This essay is essentially reproduced as pp. 160–193 of Macpherson's *Possessive Individualism* (n. 59 supra).

[87] Richard H. Tawney: "Harrington's Interpretation of His Age," Proceedings of the British Academy, 1941, **27**: 199–223, esp. p. 200.

[88] Harrington's influence on American political organization is presented in H.F. Russell Smith: *Harrington and His Oceana: A Study of a 17th Century Utopia and Its Influence in America* (Cambridge: Cambridge University Press, 1914). See also Theodore Dwight: "James Harrington and His Influence upon American Political Institutions and Political Thought," *Political Science Quarterly*, 1887, **2**: 1–44.

[89] Charles Francis Adams (ed.): *The Works of John Adams, Second President of the United States: With a Life of the Author*, vol. 4 (Boston: Charles C. Little and James Brown, 1851 – reprint, New York: AMS Press, 1971), p. 428.
[90] James Harrington: *Works: The Oceana and Other Works*, ed. John Toland, with an appendix containing more of Harrington's political writings first added by Thomas Birch in the London edition of 1737 (London: printed for T. Becket, T. Cadell, and T. Evans, 1771; reprint, Aalen [Germany]: Scientia Verlag, 1980); cited here as Toland. For a brief listing of printings and editions of Toland's collection, see Blitzer (n. 93 infra), pp. 338–339, and for a fuller account see J.G.A. Pocock (ed.): *The Political Works of James Harrington* (Cambridge/London/New York: Cambridge University Press, 1977), pp. xi–xiv; this edition by Pocock is cited here as Pocock and is used for quotations from Harrington's text. Examples of the kinds of changes which Toland made in Harrington's text are given in n. 97 infra. Of the Toland editions, I have consulted, in addition to the reprint listed above, the original collection by John Toland: *The Oceana of Iames Harrington and His Other Works* (London: Printed [by J. Darby], and are to be sold by the Booksellers of London and Westminster, 1700); *The Oceana and Other Works of Iames Harrington*, 3rd ed., with Thomas Birch's appendix of political tracts by Harrington (London: Printed for A. Millar, 1747); *The Oceana and Other Works of James Harrington* (the London edition of 1771 as noted above); and *The Oceana of James Harrington, Esq., and His Other Works*, with the addition of *Plato Redivivus* (Dublin: Printed by R. Reilly for J. Smith and W. Bruce, 1737). Adams's library contained two printings of Toland's Harrington: The London edition (3rd ed.) of 1747 and the London edition of 1771; see *Catalogue of the John Adams Library in the Public Library of the City of Boston*, ed. Lindsay Swift (Boston: published by the Trustees, 1917).
[91] *Works* (see n. 90 supra), p. xv.
[92] Pocock (n. 90 supra), p. 164; also Toland (n. 90 supra), p. 37; also James Harrington: *Oceana*, ed. S.B. Liljegren (Heidelberg: Carl Winters Universitätsbuchhandlung, 1924 – Skrifter utgivna av Vetenskaps-societeten i Lund, no. 4; reprint, Westport, Conn.: Hyperion Press, 1979), p. 15; also *Works* (n. 90 supra), p. 37. This last edition is cited as Liljegren. I have also consulted James Harrington: *The Common-Wealth of Oceana* (London: printed by J. Streater for Livewell Chapman, 1656); on this and the other "first edition," see Pocock (n. 90 supra), pp. 6–14. The text of *Oceana* and *A System of Politics* from Pocock's edition of all of Harrington's political works (1977; n. 90 supra), have been reprinted, with a new introduction, as James Harrington: *The Commonwealth of Oceana and A System of Politics*, ed. J.G.A. Pocock (Cambridge: Cambridge University Press, 1992). There is also a useful edition by Charles Blitzer of *The Political Writings of James Harrington: Representative Selections* (New York: The Liberal Arts Press, 1955).
[93] Judith N. Shklar: "Harrington, James," *International Encyclopedia of the Social Sciences*, vol. 6 (New York: The Macmillan Company & The Free Press, 1968), p. 323; Russell Smith (n. 88 supra). Charles Blitzer's *An Immortal Commonwealth: The Political Thought of James Harrington* (New Haven: Yale University Press, 1960) is the best informed and most authoritative work on Harrington and is cited as Blitzer; a convenient list of Harrington's publications is given on pp. 337–339. Also worth consulting is Charles Blitzer's doctoral thesis: "The Political Thought of James Harrington (1611–1677)" (Harvard University, 1952). A useful briefer presentation is given in Michael

Downs: *James Harrington* (Boston: Twayne Publishers, 1977). An important critical survey of interpretations of Harrington is given in J.G.A. Pocock: *Politics, Language, and Time: Essays on Political Thought and History* (Chicago: The University of Chicago Press, 1989), ch. 4, "Machiavelli, Harrington, and English Political Ideologies in the Eighteenth Century."

Harrington opposed the idea that the state should be modelled on a machine or constructed on mathematical principles. His attack was obviously directed at Hobbes, who appears in *Oceana* as an almost omnipresent target under the name "Leviathan." It has recently been argued, however, that Harrington was, to a considerable degree, a follower of the Helmontian philosophy, that he "appears Helmontian in his scorn for the use of mathematics in the 'new mechanical philosophy.'" Thus when Wren attacked Harrington for having assumed a perpetual mechanics, Harrington replied that "in the politics there is nothing mechanic or like it" and that to suppose so "is but an idiotism of some mathematician." See Wm. Craig Diamond: "Natural Philosophy in Harrington's Political Thought," *Journal of the History of Philosophy*, 1978, **16**: 387–398 esp. pp. 390, 395. Diamond argues further (e.g., p. 397) that not only was the concept of a Helmontian *spiritus* important in Harrington's philosophy of nature; "Harrington incorporated a number of related conceptions of *spiritus* within his political philosophy." Exploring Harrington's philosophy of nature from a new scholarly perspective, the author of this original and important analysis does not mention Harrington's concept of political anatomy nor does he explore Harrington's use of the science of William Harvey.

[94] Pocock (n. 90 supra), p. 656; also James Harrington: *The Art of Law-Giving: In III Books; The Third Book: Containing a Model of Popular Government* (London: Printed by J. C. for Henry Fletcher, 1659), p. 4; also Toland (n. 90 supra), p. 403.

[95] Pocock (n. 90 supra), p. 656; also *The Art of Law-Giving* (n. 94 supra), p. 4; also Toland (n. 90 supra), pp. 402–403.

[96] Pocock (n. 90 supra), p. 162; also Liljegren (n. 92 supra), p. 13; also Toland (n. 90 supra), p. 36. Cf. Harrington's *Politicaster*, in Pocock, p. 723 (and see also Toland, p. 560), where Harrington insists that "in the politics," as in anatomy, what counts is "demonstration out of nature"; politics must follow "the known course of nature."

[97] Pocock (n. 90 supra), p. 287; also Liljegren (n. 92 supra), p. 149; also Toland (n. 90 supra), p. 149. In his edition Toland has changed "store" to "matter," "gusheth" to "spouts," and "life blood" to "vital blood." That Toland and not a later editor is the author of these changes is indicated by their appearing in his edition of 1700 (n. 90 supra), p. 161. Earlier in *Oceana*, Harrington compares the Council of Trade to the Vena Porta (Pocock, p. 251; also Liljegren, p. 110; also Toland, p. 118).

[98] Judith N. Shklar: "Ideology Hunting: The Case of James Harrington," *The American Political Science Review*, 1959, **53**: 689–691.

[99] René Descartes, *Discours de la méthode*, ed. Charles Adam and Paul Taunery, vol. 6 (Paris: Léopold Cerf, Imprimeur-Éditeur, 1902; reprint, Paris: Librairie Philosophique J. Vrin, 1965), p. 47. Descartes's discussion of Harvey appears in part 5 of the *Discours de la méthode*. See René Descartes: *Treatise of Man*, French text with trans. and comm. by Thomas Steele Hall (Cambridge: Harvard University Press, 1972).

[100] Walter Pagel: *William Harvey's Biological Ideas: Selected Aspects and Historical Background* (Basel/New York: S. Karger, 1967), p. 233.

[101] See I.B. Cohen: "A Note on Harvey's 'Egg' as Pandora's 'Box,'" pp. 233–249 of Mikulás Teich & Robert Young (eds.): *Changing Perspectives in the History of Science: Essays in Honour of Joseph Needham* (London: Heinemann, 1973).
[102] See F.J. Cole: *Early Theories of Sexual Generation* (Oxford: Clarendon Press, 1930).
[103] Pocock (n. 90 supra), p. 839; also Toland (n. 90 supra), p. 470.
[104] Ibid.
[105] William Harvey: *Disputations touching the Generation of Animals*, trans. Gweneth Whitteridge (Oxford/London: Blackwell Scientific Publications, 1981), pp. 96, 99; see also William Harvey: "Anatomical Exercises in the Generation of Animals," in *The Works of William Harvey*, trans. Robert Willis (London: printed for the Sydenham Society, 1847; reprint, New York/London: Johnson Reprint Corporation, 1965 – The Sources of Science, no. 13; reprint, Philadelphia: University of Pennsylvania Press, 1989 – Classics in Medicine and Biology Series), pp. 235, 238; also William Harvey: *Anatomical Exercitations concerning the Generation of Living Creatures*, trans. (London: Printed by James Young for Octavian Pulleyn, 1653), pp. 90, 94.
[106] Whitteridge trans. (n. 105 supra), pp. 96, 101; also Willis trans. (n. 105 supra), pp. 235, 241; also 1653 trans. (n. 105 supra), pp. 89, 97.
[107] Whitteridge (n. 13 supra), p. 218.
[108] Pocock (n. 90 supra), p. 839; also Toland (n. 90 supra), p. 470.
[109] Pocock (n. 90 supra), p. 840; also Toland (n. 90 supra), p. 470.
[110] *The Prerogative of Popular Government: A Politicall Discourse in Two Books* (London: Printed for Tho. Brewster, 1658 [1657]), p. 20; also *Works* (n. 90 supra), p. 232.
[111] Pocock (n. 90 supra), p. 412; also *Prerogative* (n. 110 supra), p. 21; also Toland (n. 90 supra), p. 232.
[112] Whitteridge trans. (n. 105 supra), pp. 8–10; also Willis trans. (n. 105 supra), pp. 152–153.
[113] Whitteridge trans. (n. 105 supra), pp. 12–13; also WIllis trans. (n. 105 supra), pp. 157–158.
[114] Quoted in Kenneth D. Keele: *William Harvey: The Man, the Physician, and the Scientist* (London/Edinburgh: Nelson, 1965), p. 107.
[115] On Harvey's method see especially Walter Pagel (n. 100 supra).
[116] Pagel (see n. 100 supra), pp. 24, 331 (with qualifications, e.g., on pp. 24–25, 330–331). See also Charles Singer: *The Evolution of Anatomy* (New York: Alfred A. Knopf, 1925), pp. 174–175; Keele (n. 114 supra), p. 190.
[117] Pocock (n. 90 supra), p. 310; also Toland (n. 90 supra), p. 170.
[118] Blitzer (n. 93 supra), p. 99; Pocock (n. 90 supra), p. 723; also Toland (n. 90 supra), p. 560.
[119] Ibid.
[120] Keynes 1928 (n. 13 supra), pp. 165–166, 145; also "A Second Disquisition to John Riolan, Jun., in Which Many Objections to the Circulation of the Blood Are Refuted," trans. Robert Willis (n. 105 supra), pp. 123, 109. Whitteridge trans. (n. 13 supra), p. 7; also Willis trans. (n. 13 supra), p. 7; also Keynes 1928, p. xiii; also Keynes 1978 (n. 13 supra), p. xi.
[121] Harrington was dismayed by the fact that certain "natural philosophers" (Bishop Wilkins, for example, in his *Mathematical Magick*) wrote of machines or devices that

could either not be constructed or that could never in practice work exactly as proposed in theory; see the excellent presentation in Blitzer (n. 93 supra), pp. 90–95.

[122] Pocock (n. 90 supra), pp. 198–199; also Liljegren (n. 92 supra), p. 50; also Toland (n. 90 supra), p. 65. Cf. *Politicaster* in Pocock, p. 716; also Toland, p. 553.

[123] Keynes 1928 (n. 13 supra), p. 179; also Willis trans. (n. 120 supra), p. 132. (In *De Motu Cordis* Harvey did call "the heart of creatures" the "prince of all, the sun of their microcosm" (see n. 14 supra), but that does not mean that he favored the heliocentric system of Copernicus; cf. Whitteridge trans. (n. 13 supra), p. 76; Keynes (n. 13 supra), p. 47; Franklin trans. (n. 13 supra), p. 59. In *De Generatione Animalium*, Harvey did not compare the heart to a central sun. Rather, adopting a geocentric position (which could be Ptolemaic or Tychonic, etc.), he called the blood "the sun of the microcosm" and compared it further to "the superior luminaries, the sun and the moon," which "give life to this inferior world by their continuous circular motions." See Whitteridge translation (n. 105 supra), pp. 381–382; also Willis trans. (n. 105 supra), pp. 458–459.

[124] Keynes 1928 (n. 13 supra), p. 168; also Willis trans. (n. 120 supra), p. 124.

[125] A wholly new interpretation of Harrington's disdain for physics (mechanics) and mathematics has been suggested by Williamm Craig Diamond: "Natural Philosophy in Harrington's Political Thought," *Journal of the History of Philosophy*, 1978, **16**: 387–398. Diamond provides convincing evidence that in this regard Harrington was, to a considerable degree, a follower of the Helmontian philosophy. That is (pp. 390, 395), Harrington may have been "Helmontian in his scorn for the use of mathematics in the 'new mechanical philosophy.'" Diamond argues further (e.g., p. 397) that not only was the concept of a Helmontian *spiritus* important in Harrington's philosophy of nature; "Harrington incorporated a number of related conceptions of *spiritus* within his political philosophy." Exploring Harrington's philosophy of nature from a new scholarly perspective, the author of this original and important analysis does not, however, mention Harrington's concept of political anatomy, nor does he explore Harrington's use of the science of William Harvey in either forming a philosophy of nature or a system of political thought.

[126] Whitteridge trans. (n. 13 supra), p. 7; also Willis trans. (n. 13 supra), p. 7; Franklin trans. (n. 13 supra), p. 7; Keynes 1928 (n. 13 supra), p. xiiil; Keynes 1978 (n. 13 supra), p. xi. Harrington made other references to Harvey, even – as Robert Frank noted – using "the discovery of the circulation to argue society's need for an innovator, in this case a single legislator to lay down a plan of government"; see Robert G. Frank, Jr.: "The Image of Harvey in Commonwealth and Restoration England," pp. 103–143 of Jerome J. Bylebyl (ed.): *William Harvey and His Age: The Professional and Social Context of the Discovery of the Circulation* (Baltimore/London: The Johns Hopkins University Press, 1979), esp. p. 120. In this context Frank quotes from Harrington's *The Prerogative of Popular Government*: "*Invention* is a solitary thing. All the Physicians in the world put together invented not the circulation of the bloud, nor can invent any such thing, though in their own Art; yet this was invented by One alone, and being invented is unanimously voted and embraced by the generality of Physicians." This treatise by Harrington is included in Pocock (n. 90 supra).

[127] Petty and Graunt are exceptions in that almost all of their writings are devoted to topics in science or mathematics in relation to general polity.

3. A CONVERSATION WITH HARVEY BROOKS* ON THE SOCIAL SCIENCES, THE NATURAL SCIENCES, AND PUBLIC POLICY – CONDUCTED BY I. BERNARD COHEN

IBC: Let me ask you, to begin with, when in your career as a scientist you remember becoming aware of problems involving the social sciences.

HB: I became aware of problems in science policy that involved what I would call social issues and social values very early in my career. But I did not think of them at that time as involving the social sciences directly. I really didn't become very conscious of this aspect of the question until perhaps the '60s, especially after 1965, when I became chairman of COSPUP, the Committee on Science and Public Policy of the National Academy of Sciences. This was also a period when we had considerable interaction with Fred Harris, who at that time was a Senator from Oklahoma and who had given prominence to such problems by proposing a National Social Science foundation. I became especially conscious of the social sciences, when I testified before the Harris subcommittee concerning that proposal. I recognized that the social sciences had something to offer in the policy process which had been neglected. I would say, therefore, that it was during the early to mid '60s that I first became explicitly conscious of the social sciences as such in their relevance to science policy.

IBC: That is more or less what I expected. It seems to be generally true for most natural scientists that relations with the social sciences didn't matter very much until the '60s. Probably the only place where this direct impingement did occur before the 1960s was in the National Science Foundation, which was giving research money to certain selected social sciences. And that is

* Harvey Brooks, Professor of Technology and Public Policy Emeritus at Harvard University, was formerly Dean of Harvard's Division of Engineering Sciences and Applied Physics. Trained as a physicist, he has been a member of the President's Science Advisory Committee and the National Science Board and has also served as Chairman of the National Academy's Committee on Science and Public Policy.

relevant to a very general question which I would like to ask before we turn to the National Science Foundation. I remember one of the newly appointed officials saying to me, "I've just discovered that we give a lot of money to the soft sciences." Have you any feeling about how, generally, scientists have regarded the social sciences and to what extent they have considered them to be "soft" as distinguished from "hard" sciences?

HB: This question goes back to the debate that took place between the Kilgore proponents and the Bush proponents when the National Science Foundation was being proposed.

IBC: By Senator Harley Kilgore, Democrat from West Virginia?

HB: Yes, he favored an agency which would, among other things, include the social sciences among those basic and applied sciences to be supported by the government. As you remember, Vannevar Bush had a more traditional "pure-science" approach. He wanted to have a foundation run by natural scientists that would be limited to basic research without any consideration of immediate practical outcome.

IBC: I remember that very well because I was on the research staff that prepared Bush's 1945 report, *Science: The Endless Frontier.*

HB: Really! I didn't know that.

IBC: Yes. Rupert Maclaurin of MIT organized a kind of "Secretariat" under the direction of Henry Guerlac who was then the historian of the Radiation Lab. One of the sections that John Edsall and I prepared was the comparison of research support in the United States and in England and France. I remember many discussions about the possible inclusion of the social sciences under the umbrella of a proposed national research foundation. I.I. Rabi was vehemently opposed to having the social sciences linked to the natural sciences. They work with opinion, he used to say, and not with fact. In the end, the Bush report did not envisage a role for the social sciences, but the enabling act left the door open by using the phrase "other sciences."

HB: In all these debates there was no consensus among natural scientists on whether the social sciences should be included, but the views of the natural scientists tended to reflect the views that existed in the population at large. Conservatives, those with more or less conservative political orientation, felt that the social sciences *were* soft. That is to say, the criteria for evidence and falsifiability of theories were not the same as in the natural sciences. Moreover, I think that politically the conservatives feared that the term "science" would become tarred with the political implications of the social sciences. Some conservatives equated the social sciences, especially sociology, with socialism.

In other words, many scientists feared exactly what did happen in the controversy over the MACOS project.

IBC: What was that? Was it "Man: A Course of Study"?

HB: Yes, that's right. This was a curriculum development project of the NSF which was to be similar to the already sponsored Project Physics, which developed an important and widely used new curriculum for physics in the secondary schools. There was great controversy, as you can imagine, over the content of MACOS. For instance, Senator Proxmire, who was a constant critic of NSF, raised issues about MACOS and John Conlon, a Congressman from Arizona held that MACOS tended to undermine basic family values. The MACOS project became so controversial that it had to be abandoned. There were many people in the science community who feared that this kind of controversy could affect the whole National Science Foundation and hurt support for the natural scientists. This was probably the greatest problem. The MACOS project was very controversial because it dealt with anthropological and social issues which themselves were subjects of great dispute and controversy, unlike any discussion about principles or theories of physics. This was the sort of issue that arose at the very beginning, when it was proposed that the social sciences be included in the NSF. Nevertheless, there was no consistent view of the social sciences on the part of natural scientists.

IBC: When you and others who come from the harder sciences, so-called, think of the social sciences, what subjects come to mind? Obviously, one of them is sociology.

HB: I would be more likely to think of sociology than I would of economics. I regard economics as being a kind of thermodynamics of the social sciences in the sense that economics starts with rather oversimplified models of reality and then proceeds to make rather rigorous deductions from those models without necessarily asserting that the models are rigorous but asserting rather that the deductions are rigorous. This procedure is analogous to the methods of thermodynamics in the field of physics. The scientist working in thermodynamics begins with some very broad postulates and then explores what can be deduced mathematically from them.

I separate economics somewhat from the rest of the social sciences. In some sense economics is the model of the social sciences that most corresponds to the physical sciences, particularly to physics. Turning from economics to the rest of the social sciences, I would say that most natural scientists would arrange the rest of the social sciences in a hierarchy of softness, ranging from quantitative sociology, survey research, and so on at one end, to political science at the other, although I realize, of course, that survey research cuts across sociology and political science. But in the softer social sciences there is a characteristic verbal tradition. Political science has been a verbal, descriptive science until very recently; sociology very early became part of a quantitative tradition.

I don't believe that there was any specific image of the social scientist in the mind of natural scientists. There was, however, one point of view in the natural sciences which I think it is important to underline. That is, there is a certain degree of skepticism among natural scientists concerning the models used in the social sciences. And this skepticism was based on the scientists' own experiences in the natural sciences. The difficulty of formulating models with which to work in the natural sciences was severe enough, even though one had a relatively small number of entities which were interchangeable – like electrons, protons, and so on. In the social sciences, however, every individual is in some sense different and although individuals have certain common properties there is a degree of arbitrariness in the classification scheme. Consequently, I think there was a skepticism among natural scientists whether social scientists could make models which bore sufficient resemblance to reality so that rigorous deductions would be valid. One could not be sure something important wasn't being left out of the model even in the natural sciences, so that one could not rely on the deductions, not because they weren't rigorous but because the model did not correspond sufficiently with the real physical world.

IBC: Let us explore this a moment. My studies show that most people from the natural sciences think primarily of sociology as . . .

HB: As a social science.

IBC: . . . as *the* social science. The discussions about the scientific character of the social sciences do not usually have much to do with history, political theory, archaeology, social anthropology or social psychology. I have always found it interesting that my acquaintances who are most skeptical about the alleged scientific character of the social sciences come from the physical sciences or mathematics. Would the same be true for those in the earth science? Or in the biological sciences? When you have discussions on these topics in the various organizations to which you belong, do people from different areas have different attitudes about the social sciences? For example, some natural scientists complain that in the social sciences prediction rates are not very high.

HB: Yes, that's true.

IBC: But earth scientists have great difficulty making predictions and yet are not rejected on this account. Earth scientists would not give success in prediction the importance which physicists would.

HB: That is very true. In fact, I have made that point explicitly in an article I wrote for *Minerva* in 1972 as a commentary on Alvin Weinberg's introduction of the concept of trans-science.

IBC: I remember that. Wasn't he Director of the Oak Ridge National Laboratory and a member of PSAC, the President's Science Advisory Committee?

HB: Yes. I objected somewhat to Al's use of the word "trans-science" because he seemed to categorize the social sciences as all in the field of trans-science. One of the problems I raised was that the social sciences had been taking Newtonian mechanics as their model, whereas meteorology was a much better model. I pointed to the work of Lorenz at MIT. He had done some fundamental work on the predictability of the weather, in which he showed rigorously from the equations of motion of the atmosphere that it is impossible in principle to predict the weather more than about fifteen days ahead, no matter how perfect the information on initial boundary conditions. This situation is connected with boundary layer phenomena because it is necessary to specify the boundary conditions, in effect, with infinite accuracy. In effect, infinitesimal changes in the initial conditions or boundary values of variables lead to large changes in predictions from the rigorous equations of motion. Thus arbitrarily small errors in the boundary values of the variables lead to finite differences in outcome of solutions of the equations. Another way of putting it is that the underlying assumption of the law of causality, that effects are proportional to causes, is violated, and can be mathematically inherent in apparently deterministic equations, such as the Navier-Stokes equations of hydrodynamics. Infinitesimally small causes can lead to finite effects mathematically. You can predict that tornadoes are likely to occur in a certain broad region at a certain time, but you can't predict exactly where they will occur or what their path will be. Thus the limit to predicting the weather is really determined by the boundary phenomena. This insight was perhaps the first inkling of the theory of chaos which has become so fashionable now. In fact, some of the writers on the theory of chaos refer back to that original paper of Lorenz in the '60s.

Yet the scientific mental model which was being used by most social scientists was still the idealized model of Newtonian mechanics. They were unaware of Lorenz's result or of many other models in physical sciences based on rigorous equations, which nevertheless could not make accurate predictions. After all, the hydrodynamic equations are rigorous, but they nevertheless exhibit boundary layer and turbulent phenomena that make the problem of predictability fundamentally different from what it is in a Newtonian mechanics of billiard balls or planets. In fact, one of the first people to do significant work in this area was George Carrier, who originally became famous through his mathematical analysis of boundary layer phenomena.

IBC: As I look through the literature I find that much of social science, even economics, still tries in some way to imitate Newtonian rational mechanics, sometimes with energy physics thrown in for good measure.

HB: Yes.

IBC: Even though much of even "classical" physics does not.

HB: Right.

IBC: There has also been a large infusion of energetics, particularly in economics.

HB: You asked a question from which I digressed somewhat: What was the attitude of biologists as compared with that of other scientists? I would say that the division of biologists into systematists and molecular biologists was so great that separate answers must be given for the two halves of biology. The molecular biologists, or, more broadly, the reductionist biologists, thought more like physicists, whereas the systematic evolutionary biologists thought somewhat more like social scientists and were inclined to be more sympathetic to the methodology of the social sciences.

IBC: Perhaps you might in this connection say something about the relative importance of physical scientists as opposed to biological scientists in all of the public bodies such as the National Science Foundation, the National Academy of Sciences, and the President's Science Advisory Committee.

HB: In the original constitution of PSAC, the President's Science Advisory Committee, there was, you might say, a compromise between representing the disciplines and assembling a group of people with a common language who could communicate with each other easily in their own code. In the early days of PSAC this adjustment was made very much in terms of the latter model, with the result that PSAC was dominated by physicists in the days of Eisenhower. There were various reasons for this situation. A major one was that most of the physicists on PSAC came out of a common experience in World War II, working on various kinds of military systems. Therefore they already had a considerable familiarity with the kinds of questions and issues that were first directed at PSAC by President Eisenhower, most of which arose out of the rivalries among the military services and the proponents of various military systems. Thus, it was very natural that in its early days PSAC was dominated by physical scientists. In fact, for some time these experts did not see the need of any other forms of expertise. This was not because they were blind to other forms of expertise, but rather because of the kinds of questions that were posed.

One of the first issues that faced PSAC that involved social considerations and had very high visibility was civil defense. A crisis over civil defense arose after Kennedy's confrontation with Khrushchev over Berlin, and Khrushchev's threat to abrogate unilaterally the four-power arrangement in Berlin. There was great internal pressure in the government for a major civil defense program. The Defense Department was given the responsibility for developing that program and they conceived it almost entirely in terms of physical and engineering aspects of civil defense: bomb shelters and so on. This became a very emotional subject. PSAC had formed its own panel on civil defense to monitor and advise on implementation of the program of the Department of Defense. It soon became apparent that while, as I have said, the program of the Department of Defense was almost exclusively focused on the physical and engineering aspects, the public and the press were much more concerned with the social, behavorial, and organizational aspects. With the blessing of PSAC, Jerrold Zacharias, who was very active in many aspects of science policy questions, put together a small informal panel in Cambridge that met on weekends to discuss the psychological and behavioral aspects of civil defense planning, and particularly the rising public reaction. I was a member of this panel, which included physicists on PSAC and three or four well-known psychoanalysts and psychiatrists. Some of the people involved were Ed Purcell the physicist from Harvard; Doug Bond, head of the department of psychiatry at Case Western Reserve; Grete Bibring, a well-known Cambridge psychoanalyst; Gardner Quarton, a professor of psychology at the University of Michigan; and Oliver Cope, the MGH surgeon, who had become greatly concerned over this issue, in part because of his experience in the Coconut Grove fire in Boston in the early 1940s. I don't remember whether or not Erik Erikson, the psychoanalyst, came to any of the meetings or whether a few of us talked with him privately about these matters. We had meetings every weekend during the height of the civil defense scare, with several of the people, like Gardner Quarton and Doug Bond, actually flying to Cambridge every weekend at their own expense. This group had considerable influence in helping Jerry Wiesner, the President's Science Advisor, to persuade President Kennedy that civil defense was not just a problem of physical hard science and that it was necessary to worry especially about what would happen after the first two weeks following a nuclear exchange. The protection of the population in the first week or so after an exchange of nuclear weapons was only a minor part of the challenge compared to the social disintegration and the psychological and social problems that would ensue during a later period. This was one of the earliest involvements of social scientists in the work of PSAC. There was no formal report or conclusions resulting from these meetings, but there was considerable mutual education. Informal advice from the panel to the Science Advisor, Jerry Wiesner, helped him to convince President Kennedy to hold the national program to more modest proportions than had originally been contemplated. This situation was one of the first, I think, in which PSAC really became conscious of the psychological and social aspects of such problems. Concomitantly it was obvious that the social sciences had something important to say about many other public policy issues involving the natural sciences or technology.

IBC: Your point seems extremely important because it bears on a very fundamental question, that of the social sciences generally in relation to policy questions.

HB: Yes. The influence of the social sciences on the formulation and implementation of public policy is not a new phenomenon. During the Great Depression of the 1930s there was a strong belief that the scientific study of social phenomena could make an important contribution to the solution of the national crisis which was then the focus of political attention. At the end of World War II, in the debates over the creation of Vannevar Bush's proposed National Research Foundation, one of the principal issues was whether there would be a role for the social sciences in the new policy for post-war government support of science. William F. Ogburn, the sociologist from the University of Chicago who specialized in the study of technological innovations, testified to Congress that inventions inevitably precipitate social change and social problems and that therefore a government which supports discovery and invention has a responsibility to support the social sciences in order to help foresee and deal with the problems resulting from the discoveries and inventions growing out of its support of the natural sciences and engineering.

With the heating up of the Cold War in the period from 1949 to the early 1960s, interest in the social sciences as underpinning for public policy waned, but it revived again in the early 1960s and early 1970s with the advent of the Great Society programs and rising public awareness and concern about the side-effects of technology.

In the late '70s and in the decade of the '80s, rising concern about the decline of the international competitiveness of the United States and the alienation of the American work force led to a new dimension in the application of the insights of social sciences to public policy. This interest was stimulated by growing awareness that Japanese success in international competition was as much due to innovations in work organization and the management of the work force as to advances in production hardware. With the advent of the Reagan administration in the 1980s, however, the policy relevance of the social sciences suffered a temporary eclipse as far as government was concerned, but it is now arriving in a new form with growing appreciation of the need for a better understanding of science and technology as social systems in order to provide a basis for the formulation of more coherent science policies. This understanding is important both in considering the allocation of national resources for scientific research and technological innovation and for understanding the role of scientific knowledge and scientists in the formulation of public policies with high technical content.

IBC: You have spoken already about a specific instance involving PSAC. Can you say anything further about the acceptance of social scientists on PSAC and about the consequent influence of the social sciences on the formulation of science policy?

HB: In order to answer your question fully I have to go back into a little history. Let me start with the years of the Eisenhower administration, when there was a Science Advisory Committee which since 1951 had functioned out of the Office of Defense Mobilization. In 1957, this was transformed into the President's Science Advisory Committee under the personal patronage of President Eisenhower and quickly became known as PSAC. The occasion for this transformation was the crisis of public confidence which followed the successful Soviet launch of Sputnik in October 1957. This even raised serious concern about science and science policy, and the great step taken by Russian scientists and engineers called for a new and more important role of natural scientists in the highest councils of government. The creation of PSAC became one of the most commented upon and studied events in the evolution of science policy after World War II, not only in the United States but elsewhere in the world.

In December 1957, two months after Sputnik, Eisenhower appointed James Killian as his Science Advisor. While it is true that Killian himself was not a scientist but had been trained as a humanist and became experienced as a generalist administrator, he had worked closely with scientists, understood their attitudes, and had become highly skilled in interpreting science to laymen. I think it is important for people to understand that the first (and in some ways the most successful) presidential science advisor was not a scientist at all. He had simply absorbed the scientific culture most of his life by being around MIT and working with scientists and engineers, so that he knew how they thought and had become familiar with their language and intellectual short-cuts. It is significant that, unlike Jerry Wiesner, he almost never briefed the President by himself. He also took along one or two members of PSAC to provide the technical part of the briefing, which he could then interpret to the President in the scientist's presence.

In its original form, as I have already mentioned, PSAC itself was a remarkably coherent and close-knit group, dominated by physicists, most of whom had worked together at the MIT Radiation Laboratory or in the Manhattan Project (or both) during World War II and had been deeply involved in government advisory activities and "summer studies" closely linked to major technological events of the Cold War. Moreover, in the original membership of PSAC social scientists were notable for their absence, a fact which was not an inadvertent oversight, but was, on the contrary, an issue much debated both within PSAC and by outside critics and observers of the Committee. Among these latter were the members of a European task force appointed by the Organization for Economic Cooperation and Development, OECD, to conduct reviews or assessment of national science policy in each of the OECD member countries. This task force, two of whose three members were social scientists, deplored the lack of representation of the social and "human" sciences not only in PSAC, but also in most high level U.S. government advisory committees and boards having to do with science policy.

IBC: I know you mentioned this earlier, but could you say a little bit more in detail about the two different concepts of what PSAC ought to be.

HB: One of the elements of the debate on this subject was the existence of two quite different concepts of the role of PSAC. One view was that PSAC should be broadly representative of the range of the intellectual interests and styles of thought in the entire U.S. scientific community, and that its function should be that of a kind of ambassador from the science community to the nation's top political leadership. The other view was that PSAC should be a close-knit group of wise scientific generalists who would act as interpreters of expert scientific analyses and judgments to the President and his immediate political advisors. Both the President and the Committee opted for this second model, but many people both in the scientific public and in the larger lay public outside the government tended to take the representational view and were therefore much more inclined to look at how representative PSAC was of the scientific constituency whose interests it was thought of as, at least partially, serving. In the generalist view of the Committee's role, however, disciplinary representation would be mainly through the some 300 panelists and consultants that were tasked by PSAC or the Science Advisor to look into specific technical problems on an *ad hoc* basis for the President. PSAC could also look to the numerous committees and boards of the semi-private, semi-public National Research Council – represented on PSAC itself by its chairman, the president of the National Academy of Sciences. The some 4,000 experts serving on the National Research Council committees covered the full spectrum of the natural sciences, engineering, the social sciences, and the learned professions.

IBC: Was there any notable difference in the kind of problem addressed in the early days of PSAC and the early 1960s that might have some bearing on the involvement of social scientists?

HB: In the early days of PSAC, when it was most relied on by the President, most of the problems addressed to it arose out of military technology and involved adjudication among conflicting claims and technological proposals emanating from the three military services. These conflicts could not be resolved except at the Presidential level. The problems were highly technical, but of a type which experienced physicists could quickly learn about. PSAC, as I said before, tended to be a closely-knit club whose members spoke to each other in a condensed code that all could understand. In fact, most members, even when they came from different research disciplines, possessed a more or less common technical culture derived from similar training and exposure to some common experience, mostly with military technology. Injecting even other physicists or engineers into this culture was difficult, and indeed when PSAC did attempt to broaden its base of membership, it found that some recruits dropped out after a few meetings, reluctant to pursue the attempt to master the arcane language. To have injected social scientists into this culture at this early stage in the evolution of the committee would probably not have worked; at least that is what both sides felt at the time.

Later in the 1960s, as PSAC's agenda of problems expanded outside the national security area, a few social scientists did join PSAC – first Herbert Simon in 1968 and two years later James Coleman. Herb Simon was not appointed to PSAC as an economist but as an artificial intelligence specialist. Coleman was a sociologist. Both Simon and Coleman were strongly quantitative in their interests and had had extensive interdisciplinary experience, which made it much easier for them to penetrate the PSAC culture.

But, by and large, PSAC was very wary of the danger of having only one representative of a discipline widely different from their own. They were fearful of becoming too dependent on a single expert and not being in a position to judge among a diversity of views in a field in which they were not themselves at home. If they had to have two of every discipline, then – like Noah's Ark, – PSAC would become hopelessly unwieldy.

IBC: Could you say something about the actual day-to-day work of PSAC? How much consisted of research and developing arguments as opposed to soliciting and judging opinions from various different experts?

HB: Much of PSAC's work actually consisted of listening to arguments among experts rather distant from their own fields of specialization. In fact, they frequently acted as a sort of report review committee for draft reports of their specialized panels before they were forwarded to the President or made public. In this role they thus served as scientific generalists rather than as highly specialized experts in their own right, although they usually became quite expert in short order in the problems which they worked on. They were sophisticated enough and sufficiently familiar with scientific reasoning to ask naive but penetrating questions that forced the experts to examine their implicit assumptions and possibly revise insufficiently supported conclusions. This critical function of PSAC was never well understood by the outside world or the technical community at large.

IBC: There were, however, some essays of PSAC into the social sciences as I remember, weren't there?

HB: During the late 60s. Eventually, PSAC did branch out into the social sciences, issuing among other papers an excellent report on Youth prepared by a panel chaired by James Coleman, who, as I have said, was a member of PSAC at the time.

IBC: Was the report the one called *Youth: Transition to Adulthood* which came out in 1973?

HB: Yes. There were also several reports on issues of government support for the behavioral sciences.

IBC: One of the ones that attracted the most attention, of course, was the 1962 *Strengthening of the Behavioral Sciences.*

HB: Yes, but this was only one of several reports on aspects of support for the social sciences. For instance, there was a study of the problems of privacy and research and another on privacy and behavioral research. There were also numerous reports on education. I have already mentioned that one of the earliest involvements of social scientists in the work of PSAC never resulted in a public report. This was, of course, the work of the Civil Defense Panel, as well as the activities of the informal Cambridge group in relation to civil defense and the threat of nuclear warfare.

IBC: I believe we have said enough about PSAC. Let me turn now to the National Academy of Sciences. Would you say that the NAS furnishes another striking instance of how the social scientists as a group were only later included among the scientists and, as a further step, an example of how the social sciences were included among the sciences?

HB: The two original learned academies of the American Colonies – the American Philosophical Society of Philadelphia founded in

1743 and the American Academy of Arts and Sciences of Boston founded in 1780 – embraced all areas of learning, including the natural sciences, humanities, the arts, the learned professions, public affairs, and subjects which we would call today the social sciences, although the latter were little recognized then. This expansiveness was more in the tradition of continental Europe. When the U.S. National Academy of Sciences, or NAS, was chartered by Congress in 1863, it followed the British pattern of electing only natural scientists and a few engineers, and it continued in this pattern for its first century of existence.

IBC: This is a very interesting point. Of course there are always some exceptions. For instance, on the continent the French Academy of Sciences has been even more rigorously restricted to natural science and mathematics than the British Royal Society, but the Berlin Academy from the start included humanists and social scientists among its members along with natural scientists and mathematicians. As a matter of fact, even though the NAS was originally restricted to natural scientists, the class of natural history did include ethnology and there were almost always at least a few members who could be considered social scientists.

HB: During World War I the National Research Council, or NRC, was created for the purpose of strengthening the National Academy's advisory function to government, which has been part of its charter from its beginning, but which it has never extensively exercised. The NRC permitted the recruitment of non-Academy members to its committees, while the prestige of the NAS membership served to legitimize the advisory work of the NRC, which often dealt more with applied science and engineering than with the pure science which qualified most of the elected members. But relatively little of the NRC's work dealt with the social sciences or their application in this early period.

IBC: I have a few notes that I have put together on this subject because it is significant for our discussion. In 1865 William Dwight Whitney, the sanscritist, was a member of the NAS. In fact, he was the only member in the section called ethnology. By 1866, however, the name of this section had been changed to ethnology and philology. It also contained a second member, George Perkins Marsh, whose scholarly work certainly fit both categories. In 1899 the constitution was amended to establish six "standing committees" including one called anthropology. In 1905, as the annual report relates, the president of NAS appointed a committee which included William James "to report to the council on the relations of the Academy to the philosophical, economic, historical, and philological sciences." In that period, not only James but also Charles Sanders Peirce, the philosopher and mathematician, and Josiah Royce, the philosopher and psychologist, were members. In 1910 John Dewey was elected, and in 1911 he became a member of the new committee called anthropology and psychology. In 1965 the Academy formally recognized a class of "biological and behavioral sciences," which in 1971 was divided into two classes, including one called behavioral and social sciences. In 1975 this class had four sections: anthropology, psychology, social and political sciences, and economic sciences. Even historians of science were recognized when two prominent members of my profession, Martin J. Klein and Otto E. Neugebauer, were elected in 1977; they were classified respectively under physics and astronomy. Klein had started out as a solid state physicist and would have merited election for that earlier scientific work, but Neugebauer had never been a practicing astronomer. In 1977, when Robert Merton was chairman of political and social sciences, Kenneth Arrow was chairman of economic sciences. They had been able to change their membership category in the '70s, but when they were elected in 1968 Merton had been listed under anthropology and Arrow in applied physical and mathematical sciences. Arrow was elected as an applied mathematician and his work in this area was distinguished enough to earn him a place in the NAS but Merton was certainly not an anthropologist in the narrow professional sense and his election was based on his studies of the sociology of science.

HB: I am glad you mentioned those examples. The case of Merton and Arrow seems to me to symbolize two aspects of the history of the NAS which would repay further study. The NAS appears to have admitted representatives of disciplines other than those in the accepted categories of the natural and mathematical sciences at a time when their own fields were not yet formally recognized as sciences. These individuals, including social scientists, were included in one of the classes of members according to the prevailing system but might later change to a more appropriate class if it was introduced. It would seem, therefore, that there has been a lag between a general recognition of work as scientific – a recognition shown by election of a particular member – and a reorganization of the Academy in official recognition of the scientific character of a discipline.

More generally, it can be said that in the 20th century the natural science qualification was broadened to include a few subjects such as psychology (mostly experimental), archaeology, and anthropology (mostly physical). By the late 1960s a few of the most prestigious and quantitatively minded economists had been elected to membership. By the early 1970s new sections were added on social and political science, and thereafter the number of social scientists in the NAS grew slowly, though not in proportion to their population in the American scholarly community.

IBC: A phenomenon which seems curious to me and to many others is that certain of the social sciences such as anthropology and archaeology, the latter of which includes people engaged in biblical studies, were the first to be considered legitimate parts of the Academy of Sciences. What are the reasons that they were accepted with such relative ease?

HB: I don't know whether I can give you an authoritative answer. It is certainly true, though, that the rise of the Great Society programs increased the general visibility of social scientists as a group and helped to arouse an interest in the social sciences from a policy point of view. At the same time, attitudes towards the legitimacy of the social sciences are revealed by the fact that the elections to the Academy were highly selective, choosing those social scientists whose work was regarded as being very objective. One thinks especially of an anthropologist-archaeologist such as Bob Adams or Gordon Willey, an archaeologist elected in 1960. In general, the archaeologists were considered to be dealing with physical artifacts, and the physical anthropologists were admitted to the Academy long before social and cultural anthropologists. But even social and cultural anthropology had a legitimacy in the Academy that other social sciences did not, largely, I suspect, because they studied societies other than our own. The anthropologists who were elected to the Academy, for the most part, were those who dealt with primitive cultures. There was a concept of objectivity that seemed applicable to their work, which posited an observer of an external world – a distinction that becomes fuzzier and fuzzier as one approaches closer and closer to one's own society. Hence, the parts of the social sciences that first became legitimate were those in which the subjects or objects of study were clearly separate from our own culture. This category clearly includes both anthropology and archaeology.

IBC: I am interested in this attitude because I find that, as in France and England, in the United States the sciences in the National Academy have tended to be considered largely natural sciences in a narrow sense. The social sciences are not really a vital part. And, as we have mentioned, there is no place for humanities. And yet in the German Academy and in the Italian and Russian Academies, unlike the French Academy and the British Royal Society, there is a place for the social sciences and the humanities.

HB: This topic has another important aspect because even engineering has been very poorly represented in both the National Academy and the Royal Society, relative to their representation in the professional population.

IBC: One of the reasons may have been that the pure natural sciences in the United States at that time were struggling for recognition and were badly organized and badly supported. They didn't want to dilute their efforts by an admixture with applied sciences or engineering.

HB: Yes, that seems right.

IBC: It was hoped that the formation of a National Academy would eventually serve to promote recognition of the natural sciences. But now let us turn to other aspects of the National Academy of Sciences and to its Committee on Science and Public Policy in particular, developing the theme as you have begun to do, but also considering the way in which you personally saw that you might accomplish something for the advancement of the social sciences.

HB: Yes.

IBC: You succeeded George Kistiakowsky as chairman of the Committee on Science and Public Policy, or COSPUP, didn't you?

HB: Yes. On July first, 1965.

IBC: I take it you were the chief instigator of the famous report on the outlook and needs of the behavioral and social sciences?

HB: Herbert Simon and I were the chief instigators. Now Herb Simon was elected to the Academy before there was a class of the social sciences in the Academy, and he was elected as a psychologist because psychology did have legitimacy among the natural sciences. Consequently, Herb Simon stands out as a social scientist elected to the Academy.

IBC: Right. In 1967, when Herbert Simon was elected, there was no class or section with the designation of social science. There was a class called Biological and Behavioral Sciences which, as we have mentioned earlier, had been established in 1965, but the only sections in that class that were properly behavioral sciences were anthropology and psychology. It was not until 1971 that the class called Behavioral and Social Sciences was formally inaugurated and not until the following year that the section designated as Social, Economic, and Political Sciences was introduced. In the next year, 1972, Herbert Simon left the section of psychology for the new section, and in 1975, when this section was itself divided into two, he joined the section on social and political sciences.

HB: These details illustrate the same developments which we have seen in the cases of Robert Merton and Kenneth Arrow. But the reason why I bring up Herbert Simon's affiliation is that the Committee on Science and Public Policy, or COSPUP, was originally established with one representative from each of the sections of the Academy; at the time there were only fourteen sections, whereas now there are twenty-five. Every member of COSPUP had to be a member of a section of the Academy, so Herb Simon's election to the Academy made it possible for him to serve on COSPUP when I was chairman.

Actually, the origin of the social science report was a conspiracy between me and Herb Simon that began even before he became a member of COSPUP. Since he was chairman of the board of directors of the Social Science Research Council in 1965 and COSPUP had been doing studies of sciences such as physics and chemistry, we thought it would be a wonderful idea for the SSRC and COSPUP to cooperate in doing a study of the social sciences. Ernest R. Hilgard was asked to chair this study, and he and Herb and I and others obtained funds from the National Institutes of Health, the National Institute of Mental Health, the National Science Foundation, and the Russell Sage Foundation to do the Behavioral and Social Sciences report. We defined the subject very broadly including, for example, booklets on history and geography as well as on the more conventional social sciences. That was really the origin of the so-called BASS report, which had the official title of *The Behavioral and Social Sciences: Outlook and Needs*. It was presented as a report by the Behavioral and Social Sciences Survey Committee under the auspices of the Committee on Science and Public Policy of the National Academy of Sciences and the Committee on Problems and Policy of the Social Science Research Council. This report was published in 1969.

IBC: Do you have any feeling about what the general attitude of the Academy members was or what any particular attitudes were towards this study?

HB: COSPUP was very supportive of the idea of having a report on the behavioral and social sciences. I don't remember any skepticism at all. Although the report was produced before the Academy was reorganized to include the class of behavioral and social sciences, I have no recollection of any opposition. Rather, as I have said, everybody was very supportive of the idea of having that kind of joint study.

IBC: Did you ever receive detailed information about the effect or influence of the report?

HB: It is rather hard to quantify that. In fact, the same problem has always applied to all the COSPUP studies, and I was asked that kind of question many times when I was chairman. The recommendations of priorities, when they occurred, had almost no effect. But they did establish the agenda. Thus, all the debates that occurred were about the categories that were developed in the COSPUP reports, including those on the social sciences. In other words, what the COSPUP reports did was to develop a rather carefully selected menu of opportunities by setting forth what the content of the various subfields was and what their implications might be for practical societal problems. And that very much set the terms of debate in the political sphere about priorities in science across the board, no more and no less in the social sciences than in the natural sciences. However, it was only the priorities *within* broad fields which were debated, not between broad fields – such as between, say, physics and biology.

IBC: Who was the chief force behind the founding of COSPUP? Was it George Kistiakowsky?

HB: Yes. The origin of COSPUP was very specific. When Kistie was Science Advisor to President Eisenhower in 1959–60, one of the last things that was done before he resigned from this post at the end of the Eisenhower administration was congressional authorization of initial funding for SLAC, the Stanford linear accelerator project for research in nuclear physics, of which Pief Panofsky was director.

IBC: That was Wolfgang K.H. Panofsky, professor of physics at Stanford and a member of the President's Science Advisory Committee.

HB: Yes. Although Kistie supported the proposal, he was very much troubled by the fact that this particular project had been elevated to the Presidential level and that one of the principal protagonists was a member of PSAC. Kistie was concerned that there was no neutral body to which appeal could be made for an evaluation of this kind of proposal. He believed that the National Academy would be sufficiently removed from such specific issues that it could serve the political system in helping to make this kind of choice. That is, he held it was improper to have so many major issues dependent upon PSAC. Therefore, as soon as he stepped down from his position as Science Advisor, which he did in January 1961, he went to Detlev Bronk, then president of the National Academy of Sciences, and proposed the idea of setting up a special committee that would be unlike the National Research Council, where, as we mentioned earlier, members did not have to be members of the Academy. Bronk agreed. As constituted, COSPUP was not part of the National Research Council but was a committee composed entirely of Academy members and reporting directly to the president and council of the Academy. Its function, as envisioned by Kistie, was primarily to assess the health and opportunities of the various scientific disciplines and to do this in a public fashion which would enable its findings to enter the political debate. That was really the origin.

Very soon, however, COSPUP expanded into other fields such

as those concerned with population and various areas of public policy. The first public policy debate of this sort involved the McElroy report. William McElroy chaired a study of population which had a great deal of influence and resulted in the first federal support of family planning programs. This was called *The Growth of World Population* and appeared in 1963. From the start, therefore, COSPUP dealt with public policy problems involving science as well as with the health and opportunities of the various scientific disciplines and of sciences in general.

One of the first reports, not on the disciplines but on the general support of science, was the report on basic research and national goals which appeared in 1965.

IBC: Is that the one called *Basic Research and National Goals*?

HB: Yes. It was produced at the request of the House Committee on Science and Astronautics on the initiative of Congressman Daddario of Connecticut, who chaired the subcommittee on Science, Research, and development. When the report was finally issued in March, I was about to become chairman of COSPUP, but the report was started under the chairmanship of Kistiakowsky, who was also chairman of the ad hoc panel on basic research and national goals which prepared the report and of which I was a member. In fact, I helped write quite a bit of that report. This document was of special interest because it was prepared in response to two large questions about supporting research which were posed to the National Academy of Sciences by the Congress. Congress actually awarded a contract to the Academy to produce the report.

IBC: Does COSPUP exist still in some form?

HB: Yes, but it has been transformed, although it has had a continuous existence since the beginning. After the Engineering Academy was founded in 1964, a similar group, the Committee on Public Engineering Policy, was established by the new Academy. This happened in 1966, and the committee received the acronym COPEP. In 1981 COSPUP was reorganized as a joint committee of the National Academy of Sciences, the National Academy of Engineering, and the Institute of Medicine and was renamed the Committee on Science, Engineering, and Public Policy, or COSEPUP. Originally, every section of the Academy of Sciences was represented on COSPUP, but as the Academy expanded the number of sections the membership of COSPUP no longer was based primarily on representation from all the sections. In addition, since the members of COSEPUP are members of either the Academy of Engineering, the Institute of Medicine, or the Academy of Sciences, the committee is no longer confined to the Academy of Sciences as it was in the early '60s, when there was no Institute of Medicine or Academy of Engineering. Hence, the answer to your question is that COSPUP does in fact still exist, as COSEPUP, and is in fact very active.

IBC: But it does not still issue the older style of report?

HB: True, it stopped issuing reports on the disciplines.

IBC: Are the current activities and reports devoted to general or to specific scientific issues?

HB: At the beginning of Frank Press's regime, when Jay Keyworth was the Science Advisor, COSEPUP produced a series of ad hoc reports on what I would call new opportunities in science or opportunities for new initiatives in science. These were quite good reports.

IBC: We can keep in mind that Frank Press became president of the National Academy of Sciences on July 1, 1981, and that George A. Keyworth II was President Reagan's first Science Advisor, serving in that capacity and as Director of the Office of Science and Technology Policy, from 1981 to 1985.

HB: Yes, and that is relevant to another issue which concerned COSEPUP and had its origins in 1976. You remember that PSAC, the President's Science Advisory Committee, was abolished in 1973 and that in 1976 a law was passed by Congress to create the Office of Science and Technology Policy and the post of Director of the Office of Science and Technology Policy and Science Advisor to the President. As a consequence, a new form of PSAC was recreated in 1976, largely under the initiative of Nelson Rockefeller, who was Vice President to President Ford. Guy Stever – that is H. Guyford Stever, who was Director of the National Science Foundation between 1972 and 1976 – became the first new Science Advisor. After Ford was defeated by Jimmy Carter in 1976, there was quite a long hiatus during which Carter did nothing about appointing a Science Advisor. Finally, however, Frank Press was appointed.

It must be kept in mind that the '76 law, over the opposition of almost all past Science Advisors, mandated a rather elaborate system of reports. One of these was to be a five-year outlook for American science. Another was to be an annual report. The President was supposed to draw upon information prepared by the director of the Office of Science and Technology Policy in order to transmit a report to Congress containing recommendations for legislation and policies. Neither Frank Press nor Jimmy Carter particularly wanted these responsibilities, and there was much discussion. What finally happened was that in 1977 the responsibility for both the five-year outlook and the annual report was delegated to the National Science Foundation. In 1978 the National Science Foundation contracted with the National Academy of Sciences to do a report which would constitute a large segment of the five-year outlook. By 1982 the responsibility for developing the NAS report had been given to COSEPUP.

IBC: Was the five-year outlook published by the National Science Foundation or by the National Academy of Science?

HB: By both. For example, the first five-year outlook done by the National Academy of Sciences appeared as a separate book published by W.H. Freeman in collaboration with the NAS in 1979.

IBC: That was the one called *Science and Technology: A Five-year Outlook*?

HB: Yes. But that report was also published as a part of the NSF five-year outlook in 1980. The NSF publication was much larger, consisting of two volumes.

IBC: That was the publication entitled *The Five-Year Outlook: Problems, Opportunities and Constraints in Science and Technology*.

HB: Yes. The first volume was a topical synthesis prepared by the NSF, while the second volume comprised source materials divided into three sections, of which the first was the NAS report, the second contained reports submitted by selected government agencies, and the third consisted of papers written by individual specialists.

IBC: Was the second five-year outlook prepared and presented in the same way?

HB: Very much so. In this case, the NSF document consisted of three volumes, which appeared in 1982.

IBC: That one was called *The Five-Year Outlook on Science and Technology, 1981*.

HB: This time the NSF publication contained one volume of synthesis and generalization prepared by the NSF on the basis of the source materials published in the two "source volumes." I have these volumes right here. The second of these source volumes had three sections, of which the first was a report from the American Association for the Advancement of Science called *Policy Outlook: Science, Technology, and the Issues of the Eighties.* This was also separately published in 1982 by the Westview Press with the elements of the titled reversed to read *Science, Technology, and the Issues of the Eighties: Policy Outlook* and credited to Albert H. Teich and Ray Thornton as editors for the American Association for the Advancement of Science.

IBC: Wasn't there also a contribution from the Social Science Research Council?

HB: Yes, the second section of the second volume of source materials was a report from the SSRC entitled *The Five-Year Outlook for Science and Technology: Social and Behavioral Sciences.* I don't know whether this was published separately by the Social Science Research Council.

IBC: Brief descriptions of the essays did appear in the Social Science Research Council's *Items* and an announcement of the NSF publication also appeared in *Items.*

HB: The third section of the second NSF source volume contained "perspectives" presented by federal agencies. But what is of special interest to us now is this first volume, which consisted of the report submitted by the National Academy of Sciences. This was entitled *Outlook for Science and Technology: The Next Five Years* and was issued as *A Report from the National Research Council.* This was also published separately in 1982 by W.H. Freeman in collaboration with the National Academy of Sciences.

IBC: What about later outlooks?

HB: The third and fourth were prepared by the Committee on Science, Engineering, and Public Policy and were published in 1983 and in 1985 or 1986. These were very slim volumes.

IBC: Let's turn next to the National Board and the National Science Foundation. I believe that in many ways this may prove to be a most interesting topic because there has been a rise and fall of the social sciences, so to speak.

Let's go back to when President Truman signed legislation

creating the National Science Foundation along the basic-scientific-research lines advocated by Vannevar Bush in his 1945 report, *Science; The Endless Frontier.* In addition to the Director, the law empowered a twenty-four member National Science Board to administer the Foundation. Most of those chosen by Truman were scientists, scientists turned administrators, or industrialists and leaders of public affairs whose work related to science. Only two represented the social sciences. When did you become a member?

HB: I became a member of the National Science Board in 1962 and retired in 1974. Thus I had two full, legally permissible, six-year terms, on the Board. I think Phil Handler and I were the only two people who actually stayed on the Board for the specified maximum "twelve years." What questions should I address?

IBC: For this conversation the primary question is the role of the social sciences in the NSF. These years when you were on the National Science Board are really the crucial ones.

HB: Yes, I agree.

IBC: In 1960, the National Science Foundation established a separate Division of the Social Sciences. Ten years earlier, at the beginning of the Foundation, research support focused almost exclusively on the physical, mathematical, engineering, biological and non-clinical medical sciences. For most of the fifties, a token support of the social sciences was introduced by giving a broad interpretation of the term "and other sciences" in the enabling act. In those days the social sciences were funded under "psychobiology," "anthropology and related sciences," and "sociophysical sciences." It was not until 1958 that the Foundation formally established an Office of Social Sciences to support basic research in the anthropological, economic, and sociological sciences as well as in the history and philosophy of science. Two years later, the Office of Social Sciences was reconstituted as the Division of Social Sciences. The latter continued to function as a separate division of the NSF from 1960 to 1975.

Hank Riecken, a social psychologist, served as Head of the Office of Social Sciences. With the reorganization, he became Assistant Director for the Division of Social Sciences and continued in that capacity until 1964 when he was named Associate Director for Education.

HB: That seems right. The important thing is that social sciences had always been to some degree supported by the NSF.

IBC: Yes, but on a very limited scale.

HB: During the Great Society period, however, there was a tremendous increase in the support of the social sciences, about four fold within a few years.

IBC: That's right. Now my notes show that in 1967, the Democratic Senator from Oklahoma, Fred Harris, proposed the creation of a national social science foundation modeled on the National Science Foundation. Not only was there little support in Congress for a separate social science foundation, but social scientists themselves testified at Congressional hearings, expressing serious misgivings. Therefore, instead of authorizing a separate social science foundation, there was the 1968 Daddario-Kennedy Amendment, which modified the original National Science Foundation charter. The new legislation permitted the Foundation to sponsor applied research and designated the social sciences as a field eligible for support. How were these developments viewed by the National Science Board?

HB: I don't remember very much debate in the National Science Board about the desirability of encouraging and using the social sciences. But I do recall very well the effect of the proposal of Senator Fred Harris. You're right about the social scientists; most of them were opposed to the Harris plan. Pendleton Herring, President of the Social Science Research Council, and many others did not want to have a second foundation. They preferred to increase the status of the social sciences in the NSF and in other branches of government. I don't remember any opposition, any real resistance on the Board to the social sciences, although there was some fear that they could become overly politicized. This was always a concern on the part of the Board and of the Director of NSF, a sense that the social sciences always run the risk of producing a political sensation. Thus the Board and the NSF always tried to avoid subjects and projects that had a very strong political flavor. Even in the area of political science, there was an inclination towards opinion polling and survey research, work that was empirical and quantitative, but did not support any particular political viewpoint.

IBC: You are quite right. There was a history of support for the social sciences in the National Science Foundation almost from the start. I was very active in the Foundation in the pre-Riecken and early Riecken days. That was when NSF funded the so-called "sociophysical sciences" through the Division of Mathematical, Physical, and Engineering Sciences. The subjects we supported were primarily anthropology and archaeology, mathematical economics, sociology, and the history and philosophy of science.

HB: Do you remember when that was?

IBC: In the mid-fifties, under the benevolent guidance of Ray Seeger, a physicist. The head of this section was Harry Alpert, a sociologist, who later became Dean of the Graduate School of the University of Oregon. At that time, as I recollect, some non-scientists were added to the National Science Board.

HB: I remember Father Ted Hesburgh, President of Notre Dame. But I don't know whether he was selected for his academic expertise or as a general advisor.

IBC: Of course, he would always make an important contribution whatever his official role.

HB: There were people like him on the Board who might be called social scientists, but I don't believe they were really chosen as representatives of the social sciences.

IBC: Did you personally have any feeling about creating a separate division of National Science Foundation for the social sciences?

HB: I was very supportive of the idea of the social sciences being better represented in the National Science Foundation.

IBC: Did your time on the National Science Board overlap the demise of the social science division?

HB: No. That occurred a year after I left the Board. The social sciences were reorganized into the Division of Biological, Behavioral, and Social Sciences in 1975.

IBC: To me perhaps the most interesting aspect of that whole story is that there was never any great movement either to prevent the decline of the social sciences in NSF or, afterwards, to restore a separate division.

HB: Yes. Everybody seemed to accept it.

IBC: Let me turn to another point. There has long been a great discussion, as you know, about whether the social sciences are sciences. I have recently been examining an enormous literature on this topic, much of it written by sociologists, particularly on the question of the qualifications that define a science. Then, when this question has been answered, the authors examine whether the social sciences – and sociology in particular – can meet the qualifications. The social scientists almost all seem to agree that sociology is a science, while natural scientists, for the most part, disagree. What is your view?

HB: I have always felt that the best criterion for a science is that it must be a body of knowledge in which the theories are falsifiable. I think that falsifiability, as it is defined by Karl Popper, while it should not be pushed too far, is an important characteristic that distinguishes the sciences from other activities. For instance, I don't consider the kind of sociology that Dan Bell does, much of which I greatly admire, to be "real science" because I cannot figure out a way of proving it wrong. I have done a great deal of work in close alliance with the social sciences myself, and I must admit that I always become very uncomfortable with the fact that in much of what is written, there is frequently little consideration of alternative hypotheses. This is the feature which leads me to say that the basic criterion for science should be the falsifiability of theory. In other words, the theories and the concepts in a science ought to have the possibility of being proved or disproved by evidence and analysis.

IBC: Or at least framed in such a way.

HB: Right. Framed in such a way that they can be disproved. One can push that too far, of course.

IBC: That is right. A good example is the status of biological evolution. Popper himself did a certain amount of wavering with regard to evolution. He long believed that since evolution was not falsifiable, it could not be a scientific theory. Later he changed his mind.

HB: Of course, as Bernie Davis frequently points out, evolution never became a completely solid theory until molecular biology was developed, because molecular biology provided the first real microscopic evidence of the connectedness of all life. Evolution as formulated by Darwin is somewhat analogous to thermodynamics in physics. Darwin, it seems to me, did not penetrate far enough into fundamental mechanisms that one could be absolutely sure that things could not be explained by alternative hypotheses. That is the trouble, by the way, with Popper and the Popperian criterion. Even though Popper's criterion is fundamentally correct, if one adopted it with respect to every theory, one would never make any progress.

IBC: I agree. Let us now return to the specific subject of this book: the relations between the social sciences and the natural sciences. We have thus far covered a number of aspects of this general topic. But there still remain some aspects of the public policy issue that I want to explore with you. They can be posed in relation to the famous Coleman Report of 1966. This report seems to be remarkable not only because it was a social science report that directly affected policy but also because it was actually mandated by the Congress. I believe it may even have been the first report in the social sciences ever to be mandated directly by Congress. Usually, such reports have been produced at the request of some agency in the executive branch of government.

HB: I believe that is correct.

IBC: I have just been rereading the Coleman report and also Coleman's most recent book, *Foundations of Social Theory*, published by Harvard University Press. To me what is most interesting about this huge book is that it contains very little discussion of social progress, social problems, or social policy. It looks like a book on statistical thermodynamics. This aspect is really quite extraordinary. If one examines comparable works of other sociologists – Will Ogburn's, to take a classic example, or the Hoover Report – one finds almost always a dual concern: making a social-science analysis and also an attempt to influence policy and improve the state of society, or at least to call attention to some of society's problems. Clearly there is a difference here between sociology and the physical sciences, where the primary aim is to know or to understand nature or even to control nature, but not to improve it.

HB: That would be engineering, an application of knowledge.

IBC: Good! Let's make a distinction between physics and engineering. This leads me to a fundamental question: Do you think there is a lack of this distinction in social science, notably in sociology? What I want to explore is whether the confusion between social "science" and social "engineering" may have been a factor in considerations of whether social sciences are sciences? Has it ever been a factor in the use or non-use of the social sciences in questions of public policy? In other words, do you believe that the complaints of social scientists about the very small role of the social sciences in relation to policy questions is related to a failure of social scientists to keep separate questions of knowledge and questions of advocacy? Is there any concern on the part of natural scientists that perhaps sociologists may be more concerned with the word "social" than with the word "science"?

HB: It depends. I don't think the scientific community is monolithic in this respect. It seems to me that there are two kinds of social scientist, both of which have valid claims. Jim Coleman represents one sort of extreme. From what you say about his book, he seems to have moved even further in that direction, namely, believing that it is the job of the social scientists to describe reality and let the chips fall where they may. But there are surely scientists who find that the social sciences are more like engineering. In fact there is sometimes almost an equating of social science with social engineering.

I remember one incident which made me very uncomfortable when I was on the National Science Board, during the heyday of the popularity of the social sciences in Congress. Some of the members of the Board appeared before the Congress and testified almost to the effect that the knowledge being gained in the social sciences that the NSF was supporting would enable us to engineer society. I am surprised that this attitude did not provoke a stronger negative reaction from the Congressmen. They have been said, after all, to consider social engineering their exclusive province. But I remember this event distinctly because I was there when the testimony was given. It made me very uncomfortable because it bordered on the claim that the social sciences would enable us to manipulate society. Certainly during the 1960s, especially the late 1960s, there was a period of hubris in the social sciences, when quite a few people believed that the social sciences were really going to make it possible to engineer society. And while most good social scientists denied any such claim, there was much rhetoric that suggested otherwise.

IBC: I think that this is a very important point. As I mentioned in my introduction to this book, the very first issue of the *American Sociological Review* stressed a dual aim: that it would report progress in the science of understanding society but also deal with social problems. And this social ethics or social engineering feature has grown up, side by side, with the advancing knowledge of social phenomenon. There is an ambivalence of aims here which may be similar to the situation which obtains, to some degree, in medicine.

HB: That is right.

IBC: You once told me that you believe that probably the most important of the social science reports that have affected American public policy was the Gunnar Myrdal report on *The American Dilemma: The Negro Problem and Modern Democracy.*

HB: Right.

IBC: Could you explain? I agree with you, but I find on reexamination of this report that it was not issued as a scientific study. There are no claims that it presents the latest findings of science. It became – to a large degree – a statement of Myrdal's personal convictions, even though he drew extensively on the research done over many years by a committee of research social scientists. The case was amply documented, to be sure, but it reads somewhat as an extended moral tract, as an indictment of what society had done to one particular class and its members. The Coleman Report, on the other hand, was based on a scientific investigation and issued as a scientific report. The conclusions were statistical rather than moral. The volume has the appearance of an engineering study and is not at all like a moral tract. Have you any feelings about the relative impact of the two reports? In the end, in terms of net effect on public policy, was there much difference? In other words, what is the effect of the "science" content of social science on questions of policy?

HB: That is a very difficult but important question. The Coleman Report certainly had a great impact at the time, although in retrospect some people feel that some aspects of that impact were perhaps not legitimate. Was it not the Coleman Report which indicated that there had been very little benefit from Head Start, for example? But the general tone of the Coleman Report, or the general conclusion which most people drew from the Coleman Report was that the evidence for the effectiveness of intervention in the educational systems was not very persuasive. For example, one of the things that Coleman found was that there was virtually no correlation between per-pupil expenditures in schools and academic performance.

There is, as you say, a tradition in the social sciences of both knowledge and programmed action. This reflection leads to a general consideration of the relationship between policy and science. One of the roles of science in policy is to project, as objectively as possible, the most probable outcomes of various alternative policies. There are some people who would say that this should be the only contribution of scientists to the policy process: not to recommend policies, but to predict, as well as possible from the data available, what the probable outcomes will be from various alternative policies. And this is the tradition in which Jim Coleman was operating. I think myself that we need both traditions, but people should be more candid in stating which tradition they are following.

I like to draw a parallel between the physical sciences and engineering and have often said that in the whole "science and society" field there are really two different traditions, both of which are very important and both of which have a place. One is the "science, technology, and society" tradition in which the student is primarily interested in the phenomenology of the interactions between science and technology and society. The other is the "science policy" tradition, which is more like engineering, and in which the student is more concerned with policy design. Both traditions, as I say, have a place, and they are obviously interrelated by the common effort to predict the consequences of various policies. This is also true of engineering. In engineering one uses science to predict the performance of various designs, and engineering cannot exist without science in this sense, the ability to predict the performance of various designs. It is impossible to have science policy without some ability to predict the consequences of various policies. I think, therefore, that both of these traditions have an important place in the system, and this is true whether one is talking about the natural sciences or the social sciences.

IBC: That is a very important point. I have on my shelf at the moment a continual sequence of books – there must be forty or so – which have come out, and still are coming out, all having the same lament: that applied social sciences is not as yet considered to be policy research, in the sense that it does not influence policy. There is concern that social science research or policy research is used only when the person who is determining the policy finds that the results of the research can support an already adopted position. There is thus a continuing uneasiness on the part of social scientists because they find that policy goals are not determined by their research. Would you like to say something about this?

HB: My comment on that must revert to the analogy about the relations between science and engineering. The fact is that science does not determine the design of artifacts. When a man is sent to the moon, that is not a scientific project, but it cannot be done without the ability to predict how the various artifacts that are built are going to perform. And if the artifacts are very large and complicated, it may not be possible simply to build and test them, but it will be necessary to infer their performance from analysis and from testing the components and so on. A model of this kind is given by the 1975 Rasmussen Report, published by the Nuclear Regulatory Commission under the title, *Reactor Safety Study: An Assessment of Accident Risks in U.S. Commercial Nuclear Power Plants*. In this report the system of probabilistic risk analysis is used. If the risks in a very complex system are examined, the failure rates of various components can be empirically tested, but it may not be possible to test the whole system. That can be done only by analysis, which always involved some assumptions which cannot be fully tested empirically about the statistical independence of a sequence of events. Probabilistic risk assessment always involves some incompletely testable assumptions regarding the statistical independence of individual events in a sequence leading to catastrophe. If they are truly independent, then the probabilities simply compound by multiplication and are usually very small because several events in sequence are involved in an accident with consequence. You can estimate with considerable confidence the probabilities of individual events from field experience with the failure of components (including the effects of human error), but the possibility of coincidental failures that are causally related can only be estimated through the exercise of imagination and perhaps some partial experimental testing. But you have to imagine the possible event first before you can test it, and that is why PRA can never be completely "scientific."

The expectation that policy analysis will influence policy is a somewhat different problem. People do not like to be told that their policies will not work. The problem is that the prediction of outcomes is only probabilistic and therefore much slippage can occur. The most probable outcome may not be the one that actually occurs. You can seldom prove that a policy just will not work. You can only say that it seems more plausible than not that it will not work or *vice versa*. Maybe the lament of social scientists about policy is justified.

Martin Greenberger's book, *Caught Unawares*, provides one of the best analyses I know of the effects of energy policy analysis on energy policy. One of the conclusions that Greenberger reaches is that the political effectiveness of energy reports was inversely proportional to their scholarly quality as judged by scholars. He considered the report of the highest quality to be one of several that were financed by the Ford Foundation, chaired by Hans Landsberg, and published in 1979 as *Energy, The Next Twenty Years*. It came out the same year as the CONAES study, *Energy in Transition, 1985–2010, Final Report of the Committee on Nuclear and Alternative Energy Systems*. The Ford Foundation report was given an extremely high rating by experts, but it sank without a trace almost immediately. One of the most influential reports, or temporarily influential reports, issued at the same time was the one by Stobaugh, Yergin and some other people in the Harvard Business School, which had an enormous impact and was given a low rating by scholars in Greenberger's survey. The CONAES Study was somewhere in between.

It is difficult, therefore, to give a simple answer to your question. It certainly seems to be the case that the influence of a report depends on the style and manner of presentation. In a certain sense, the more hard evidence is required to support a policy question, the more difficult it is to present a policy conclusion in a way that is accessible to a large number of people. And that is certainly a major part of the problem.

IBC: Is part of the problem related to your analogy of physics and engineering? What you say is that it is the engineering that sets the goal, not the physics. It is perhaps an idle hope to expect that the people who determine political or social policy will turn to social scientists at the stage of deciding on the goals rather than for implementation of a policy to achieve those goals.

HB: Well, perestroika would never have resulted from social science analysis. Although analysis may help to decide whether a policy design – *after* somebody has thought it up – will achieve the goals it aims at, there is a synthetic and imaginative quality to policy that is simply unrelated to analysis. The design of a policy is the result of an imaginative product, more like a work of art or the product of a craftsman, even though there is a component of analysis. There is a very high element of craft in policy design, no matter what field of policy one is discussing. And much of the debate is about how to relate the craft to the analysis. Surely analysis ought to be able to help evaluate the policies that are crafted, but there is always the role of imagination in policy design which simply does not necessarily follow directly out of the analysis. I am afraid that I have not given you a very good answer to your question.

IBC: There is no easy answer. That itself is a very interesting conclusion.

A NOTE ON "SOCIAL SCIENCE" AND ON "NATURAL SCIENCE"

Throughout certain parts of this book, the terms "natural science" and "social science" (or "natural sciences" and "social sciences") are used to designate, respectively, the physical and biological (and earth) sciences plus mathematics and the subjects known today as social or behavioral sciences.[1] Roughly speaking, these divisions correspond to the German "Naturwissenschaften" and "Sozialwissenschaften"[2] and are in current use in the Anglo-American world. The use of these two terms – natural sciences and social sciences – when dealing with any chronological period before the mid-nineteenth century is somewhat anachronistic to the degree that it imposes on earlier thought the rigid categories and values of a later time. Today the phrase "science of society" would suggest a subject much like physics or biology but in the eighteenth century and well into the nineteenth the implication would have been only a system of organized knowledge. When Thomas B. Macaulay wrote that "the science of government is an experimental science," he meant that this subject was a system of organized knowledge that was based on experience, the same sense in which these words "experimental" and "science" had been used by Hume and Burke (see Chapter 1, §1.1). Such examples alert us to the dangers of using such terms as "science" or "experimental" anachronistically.

The reader will note that throughout this volume the physical and biological sciences are in many instances referred to as "natural sciences", a term that may embrace mathematics. In an earlier presentation of my researches into the interactions of the natural sciences and the social sciences — at a meeting convened by Karl Deutsch and John Platt at the Wissenschaftszentrum in Berlin in 1982 — I introduced the dichotomy of "mathematics and the natural and exact sciences" and the "social sciences", but for convenience of discourse I abbreviated "mathematics and the natural and exact sciences" into the simpler expression "sciences".[3] In the first comment on my paper, Alex Inkeles criticized this usage. I had "obviously," he argued, implied a difference in values assigned to the two fields of endeavor, one being "science" — "natural" and "exact" — the other "social". The justice of his criticism has led me to use the term "natural science" (and its plural "natural sciences") in order to avoid any pejorative implications, even though there may be some possible ambiguity because "natural science" may wrongly suggest "natural history" or the life sciences. I have long believed, however, that if one were seeking an antonym for "natural" science, it would not be "social" science but rather "unnatural" science; which, in turn, suggests that the proper anytonym for "social" science would be "anti-social" science.

The designation "social science" arose and became current in the late eighteenth century. The introduction of "social science" has two somewhat distinct aspects. First of all there is the actual occurrence of the term; second, the emergence of a concept in which knowledge of society is perceived to be a "science" in the sense of the physical and biological sciences. A good part of this book is devoted to an exploration of the ways in which what we would call the social sciences made use of the established natural sciences, beginning with the age of the Scientific Revolution (see Chapter 2). Many examples show the different ways in which a variety of thinkers, under whatever name or rubric they classified their activity, conceived their own subject in relation to the natural sciences and mathematics of their day. Therefore, for expository purposes I may have somewhat anachronously used the term "social sciences" (and also "moral sciences") for their thoughts and writings on such topics as political theory or statecraft, organization of the state or of society, natural law, international law, economics, and kindred subjects.

I do not know who first used the terms "social science" and "science of society." In a letter to John Jebb, written from London on 10 September 1785, the American statesman John Adams (later to become the second president of the United States) wrote of "the social science." A year before, in a letter to A.M. Cérisier, he applauded the way in which French savants (Cérisier among them) had "turned to the subject of government"; he voiced his judgment that "the science of society is much behind other arts and sciences, trades and manufactures." Even earlier, in June of 1782, Adams had declared that "politics are the divine science."[4]

I do not believe that Adams invented these expressions. In those days, however, as has been mentioned, the term "science" did not have the identical meaning which it was to acquire later in the nineteenth century. The nearest equivalent of what we would consider to be a science, in the sense of a natural science, was natural philosophy, but that subject was more akin to our physics plus astronomy and part of chemistry. (See, on this topic, Chapter 1, §1.1.)

The earliest recorded use in print of the actual expression "social science" ("science sociale") seems – according to Keith Baker – to have been in 1781 in a pamphlet addressed to Condorcet.[5] It has been suggested that since the term "art sociale" was commonly used by the Physiocrats before the Revolution, perhaps the transformation to "science sociale" occurred before 1791.[6] In any event, Condorcet himself used the new term in a draft plan presented to the Committee on Public Instruction of the Legislative Assembly in January 1792. Condorcet also introduced "social science" in his writings after 1792, notably in his "Esquisse,"[7] translated under the title *Outlines of an Historical View of the Progress of the Human Mind* (London, 1795). Faced with a new and difficult expression, the British translator chose to render "science sociale" as "moral science,"[8] a name used widely in England throughout the nineteenth century for social science.[9] In France the equivalent, "sciences morales," was in common usage early in the nineteenth century, as in the name of a "class" in the Institut de France, constituted after the Revolution: Sciences Morales et Politiques.

"Social science" entered American English in a translation of Destutt de Tracy's *Treatise on Political Economy* (Georgetown, [Washington] D.C., 1817), sponsored by Thomas Jefferson, to whom Destutt had sent the manuscript, which he could not then publish in France. Jefferson apparently checked the translation and wrote a prospectus approving the use of a number of neologisms, among them "social science."[10] In British English, "social science" seems to have come into being through a circuitous route that included a Spanish translation, made by Toribio Nuñez (Salamanca, 1820), of some selections from the writings of Jeremy Bentham. Nuñez introduced "ciencia social" into the title: *Espíritu de Bentham: Sistéma de la ciencia social*. Bentham later congratulated Nuñez for his use of "ciencia social," referring to "the science so aptly styled by you the *social science*."[11]

The history of this development has been admirably encapsulated by Victor Branford as follows:

Between Vico's 'New Science' and Comte's 'Sociology' the infiltration of various kindred phrases, such as Social Science, Science of Society (Condorcet), Science of Man (St. Simon), would seem to mark a general tendency toward the expansion of science into the field of humanistic studies. Among Comte's contemporaries, J.S. Mill (only eight years younger than Comte) held pronouncedly that the time was ripe for marking off from other studies – both scientific and philosophical – a general social science, and for this he himself proposed a particular designation. In 1836 Mill defined the scope and character of this department of studies, using as titular synonyms, these, among others phrases – Social Philosophy, Social Science, Natural History of Society, Speculative Politics, and Social Economy. This essay of Mill ('On the Definition and Method of Political Economy') appeared six years before the completion of the 'Positive Philosophy.' Lacking the large historical interests of Comte, Mill necessarily conceived of Social Science in a considerably different way from Comte. But after the appearance of the 'Positive Philosophy,' Mill was very considerably modified in his views of Social Science.[12]

The use of "moral sciences" became quite extensive during the nineteenth century in England. Thus in John Stuart Mill's *A System of Logic, Ratiocinative and Inductive* (London, 1843), Book Six on "The Logic of the Moral Sciences" discusses the methodology suitable for the social sciences. But in the text itself, Mill uses both "sociology" and "the social science" as distinct from political science or political economy or history. In the beginning portion of Chapter Nine, Mill originally wrote in his manuscript about "the Social Science . . . which I shall henceforth, with M. Comte, designate by the more compact term Sociology." On reflection, however, he would not so easily pass over this neologism, based on the compounding of a Latin and a Greek root, and so the published version discusses "the Social science . . . which, by a convenient barbarism, has been termed Sociology."[13] By the end of the nineteenth century moral sciences had become the name used in Cambridge University and elsewhere for the subject now known as philosophy.

In French culture the expression "sciences morales," which had been in regular use since early in the nineteenth century, has become obsolete. Curiously enough, it has been said – by Etiemble, the quixotic defender of the purity of the French language – that the factor causing a change from "sciences morales" to "sciences humaines" was an obsession for "la classification yanquie." That is, he considers "sciences humaines" to be a new term introduced as the French equivalent of the supposedly American "social science," a name under which (according to Etiemble) "the Americans assemble history, human geography, normal and pathological psychology, and the different branches of sociology" (but not, it would appear, economics, anthropology, or political science). The editors of Dupré's *Encyclopédie du bon français* (1972) observe that the name "sciences humaines" is perhaps maladroit, since it does not include human anatomy and physiology. "Faute de mieux," they conclude, the new name should be adopted, even though "sciences morales" would "be more logical," although antiquated and even "reactionary."[14]

In Germany , as I have mentioned, the usual distinction is between "Naturwissenschaften" (natural sciences) and "Socialwissenschaften" (social sciences), but in the late nineteenth and early twentieth centuries, there came into general usage an additional distinction, "Naturwissenschaften" and "Geisteswissenschaften," roughly the natural sciences (including mathematics) and the sciences of man or, possibly, the arts and humanities plus the social sciences.[15] Current German usage also includes "Soziologie" and even "Sociologie."[16]

* * *

The use of the term "social science," as opposed to "social sciences," reflects the historical climate of the late eighteenth century and of much of the nineteenth. The emerging subdisciplines which we know as economics or sociology or political science (as opposed to political theory or political history) could then be still considered as part of a general "social science."

In America in the nineteenth century, belief in such a general subject – coupled with the goal of improving society – found expression in a strong Social Science movement which had as its stated aim "to create a special and unified science of human society and human welfare."[17] This Social Science movement has been described as "a non-political attempt to produce a social theory and a methodology which could be used as an intellectual instrument for the betterment of the lot of mankind."[18] Eventually (in 1865) there was formed the American Association for the Promotion of Social Science, on the model of the British Social Science Association and obviously patterning its name on the American Association for the Advancement of Science. In the 1880s specialized sub-disciplines broke away from the parent organization with the formation of the American Historical Society and the American Economic Association, followed by a separate organization of the political scientists. In 1909 the rise of the separate disciplines brought the general association for Social Science to an end.[19]

Another attempt in America to have a single "umbrella" organization for all the social sciences produced the Social Sciences Research Council. The SSRC differed from the older Social Science Association in that it did not set forth an ideal of a unified and general social science, but was created as a cooperative organization of separate and individual social sciences. Traditionally, the social sciences have included five fundamental disciplines: anthropology, economics, political science, psychology, and sociology. When the Social Science Research Council was organized in 1923 as the counterpart of the National Research Council, the core membership consisted of the professional or scholarly associations representing these five disciplines plus two others – history and statistics.[20] History is sometimes classed with the social sciences, sometimes with the humanities.[21] George Homans's list of "social sciences" includes "psychology, anthropology, sociology, economics, political science, history and probably linguistics."[22]

The first article in the *Encyclopaedia of the Social Sciences* (1932), written by the editor, Edwin R.A. Seligman, posits three classes of social sciences – the "purely social sciences" (the earliest ones, in historical order – politics, economics, history, jurisprudence: and the later ones, in historical order – anthropology, penology, sociology, and social work); the "semi-social sciences" (ethics, education, philosophy, psychology); and the "sciences with social implications" (biology, geography, medicine, linguistics, and art). In the Introduction to the successor *International Encyclopedia of the Social Sciences* (1968), the editor, David L. Sills, acknowledges (pp. xxi–xxii) that no final answer can be given to the question, "What are the social sciences?" The reason is that the scope of the social sciences varies from one time period to another. Sills calls attention to certain controversies, e.g., whether history is a social science or part of the humanities, whether psychology is a social or a natural science. The editors, he reports, determined that "the majority of the topical articles" would be devoted to anthropology, economics, geography, history, law, political science, psychiatry, psychology, sociology, and statistics.

Another grouping of disciplines is the "behavioral sciences," a name which came into general use in the 1950s. A major factor in the spread and acceptance of this term was its use by the Ford Foundation in a large-scale and well funded program that was at first unofficially and later officially known as "behavioral sciences." The behavioral sciences, according to Bernard Berelson, is a rubric usually understood to include "sociology; anthropology (minus archeology, technical linguistics, and most of physical anthropology); psychology (minus physiological psychology); and the behavioral aspects of biology, economics, geography, law, psychiatry, and political science."[23]

In *The Behavioral and Social Sciences* (1969), the primary subject areas considered were: anthropology, economics, geography, history, linguistics, political science, psychiatry, psychology, sociology, and aspects of mathematics, statistics, and computation.[24] This may be contrasted with *Knowledge into Action* (1969), where it is said that "historically," five social science have been "central": anthropology, economics, political science, psychology, and sociology. Other disciplines dealing with "social phenomena" are said to be demography, history, human geography, linguistics and social statistics.[25]

In the chapters of our present book, particular social sciences (e.g., economics, sociology) are referred to under their specific names while the terms "social science" or "social sciences" are used either in the nonspecific sense of former times (to include all the "sciences" relating to human behavior and to human societies) or to indicate an all-encompassing "science" that might embrace all human social activities. For the earliest periods under consideration (e.g., the Scientific Revolution in Chapter 2), theories of government or of the state (the works of Hobbes and Harrington) and the conduct of international relations (Grotius) are included under the rubric of "social sciences" because they represent areas of study which later became part of the recognized social sciences.

SOZIALWISSENSCHAFT AND GEISTESWISSENSCHAFTEN

In the twentieth century, the words "Sozialwissenschaft" and "Gesellschaftswissenschaft" can be used for sociology and also for social science. Sometimes "Gesellschaftslehre" or "Soziologie" is used as the direct equivalent of sociology. In the latter nineteenth century, however, there came into general usage a distinction between "Naturwissenschaften" and "Geisteswissenschaften," understood to encompass respectively the natural sciences (including mathematics) and the human sciences (the social sciences and the humanities).[26] Some thinkers and scholars, such as Wilhelm Dilthey in 1883 and Erich Rothacker in 1926, have suggested that "Geisteswissenschaften" owes its invention or at least its diffusion to J. Schiel, who in 1849 used this term for "moral sciences" in his German version of John Stuart Mill's *System of Logic*.[27] In rendering the title of Book VI, "On the Logic of the Moral Sciences," Schiel does write, "Von der Logik der Geisteswissenschaften oder moralischen Wissenschaften," and he generally employs "Geisteswissenschaften" for "moral sciences" in the text.[28] But the appearance of "Geisteswissenschaften" in the translation of Mill's *Logic* in 1849 seems not to have established this usage as definitive since the term is not similarly employed in the later translation of Mill's *Logic* by Theodor Gomperz, who is 1873 rendered the title of Book VI as "Von der Logik der moralischen Wissenschaften" and uses this equivalent in his text.[29] Moreover, Alwin Diemer has shown that "Geisteswissenschaft" was used as early as 1787, that "Geisteswissenschaften" is found in something like its modern acceptation in 1824, and that the modern sense is clearly attested in the distinction made by E.A.E. Calinich in 1847 between the "naturwissenschaftlichen und der geisteswissenschaftlichen Methode."[30]

The Hegelians regarded "Geisteswissenschaft" as "philosophy of spirit" and therefore as a noun in the singular. The term "Geisteswissenschaften" in the plural seems to have come into general usage as part of the development of the idea of "Geisteswissenschaften" as a set of interrelated but independent disciplines. An academic address given by Hermann von Helmholz in 1862 is of particular interest because of the author's eminent contributions to several of the natural sciences combined with his work on philosophy and fine arts. In his address, Helmholz discussed at some length various relations among "Naturwissenschaften" and "Geisteswissenschaften," indicating both their differences and their interconnections.[31] But it is Wilhelm Dilthey who should probably be considered the major figure both in the development of the concept and in the dissemination of the term "Geisteswissenschaften."[32] For Dilthey's term the English rendition until recently tended to be "human studies" but is now increasingly "human sciences."[33] Today "Geisteswissenschaften" may be considered more or less the equivalent of "human sciences" or "sciences of man" (and so somewhat similar to the French "sciences de l'homme" or "sciences humaines"), a rubric that embraces the traditional subjects of philosophy, philology, literary study, jurisprudence, history, and political science, along with the newer subjects of anthropology, archeology, psychology, economics, and sociology. Other fields, such as theology and education, may also be included with prominent subdivisions, such as the study of folklore and the history of art, even being regarded as separate disciplines.

NOTES

[1] There is no universal agreement today on which subjects of knowledge or inquiry should be included among the social or behavioral sciences; see ch. 1, §1 supra.
[2] For the additional problem of "Geisteswissenschaften" see nn. 26–28 infra.
[3] Karl W. Deutsch, Andrei S. Markovits, & John Platt (eds.): *Advances in the Social Sciences, 1900–1980: What, Who, Where, How?* (Lanham [Maryland]/New York/London: University Press of America: Cambridge [Mass.]: Abt Associates, 1986), pp. 149–253.
[4] Charles Francis Adams (ed.): *The Works of John Adams*, vol. 9 (Boston: Little, Brown and Company, 1854), pp. 512, 523, 450.
[5] Keith Michael Baker: *Condorcet: From Natural Philosophy to Social Mathematics* (Chicago: The University of Chicago Press, 1975), Appendix B: "A Note of the Early Uses of the Term 'Social Science.'"
[6] Ibid., p. 391.
[7] Baker (op. cit., ch. 4, esp. pp. 197–202) gives an excellent and succinct presentation of Condorcet's views of the "science sociale." On p. 201 Baker discusses Condorcet's concept of "social science" and indicates how Condorcet contrasted Greek political theory ("a science of facts, an empirical science, as it were") with "a true theory founded on general principles which are drawn from nature and acknowledged by reason." In the course of this elaboration, the term "political sciences" was, not surprisingly, introduced by Condorect along with "social sciences".
[8] Ibid., p. 392. For the reverse situation, in which the German translator of Mill's *Logic* introduced "Geisteswissenschaften" as the German equivalent of "moral sciences"; see the second part of this "Note of 'Social Science.'"
[9] For example, the economist Francis Ysidro Edgeworth wrote in 1881 of economics as one of the "moral sciences"; in the same work he also wrote of social science, using the French term "mécanique sociale," which he hoped would some day "take her place" as the equal of Laplace's "mécanique céleste." See Francis Ysidro Edgeworth: *Mathematical Psychics: an Essay on the Application of Mathematics to the Moral Sciences* (London: C. Kegan Paul & Co., 1881).
[10] Gilbert Chinard: *Jefferson et les idéologues* (Baltimore/Paris: The Johns Hopkins Press; Paris, Les Presses Universitaires de France, 1925), p. 43–44; also Baker (n. 5 supra), pp. 393–394.
[11] J.H. Burns: *Jeremy Bentham and University College* (London: University of London, Athlone Press, 1962), pp. 7–8.
[12] Victor Branford: "On the Origin and Use of the Word Sociology," in *Sociological Papers* (London: Macmillan and Co., 1905), pp. 5–6 quoted in L.L. Bernard & J. Bernard *Origins of American Sociology: The Social Science Movement in the United States* (New York: Thomas Y. Crowell Company, 1943), p. 3.
[13] John Stuart Mill: *A System of Logic, Ratiocinative and Inductive*, 2 vols., ed. J.M. Robson (Toronto: University of Toronto Press; London: Routledge & Kegan Paul, 1974 – Collected Works, vols. 7–8), p. 895.

It should not be thought that Comte composed this hybrid in ignorance, since he was fully aware that he was compounding a mixture of a Greek and a Latin root. But he saw no other way of having the new science define its subject to be society (using the root "*socio-*" from the Latin noun "*socius*") and also declare its stature as a science by a similarity in its final root to such sciences as biology, geology, physiology, mineralogy, and so on.

[14] Fernand Keller & Jean Batany (eds.): *Encyclopédie du bon français dans l'usage contemporain*, vol. 3 (Paris: Editions de Trévise, 1972), p. 2344.

[15] The complex history of the use of "Geisteswissenschaften" is discussed below in the second part of the present Note.

[16] See further L.H. Adolph Geck: "Über das Eindringen des Wortes 'sozial' in die Deutsche Sprache," *Sozial Welt*, 1962, **12**: 305–339.

[17] L.L. Bernard & Jessie Bernard: *Origins of American Sociology: the Social Science Movement in the United States* (New York: Thomas Y. Crowell Company, 1943), p. 3.

[18] Ibid., p. 4.

[19] Ibid., ch. 8.

[20] Ibid., p. 546.

[21] Ibid., p. 658.

[22] George Homans: *The Nature of Social Science* (New York: Harcourt, Brace & World, 1967), p. 3.

[23] Bernard Berelson, "Behavioral Sciences," *International Encyclopedia of the Social Sciences*, vol. 2 (1968), pp. 41–42. See, further, Herbert J. Spiro: "Critique of Behavioralism in Political Science," pp. 314–327 of Klaus von Peyme: *Theory and Politics, Theorie und Politik, Festschrift zum 70. Geburtstag für Carl Joachim Friedrich* (The Hague: Martinus Nijhoff, 1971).

[24] *The Behavioral and Social Sciences: Outlook and Needs* (Washington; National Academy of Sciences, 1969), a report of the Behavioral and Social Sciences Committee (operating under the joint auspices of the National Academy of Sciences and the Social Science Research Council), pp. xi, 19.

[25] *Knowledge into Action: Improving the Nations's Use of the Social Sciences* (Washington: National Science Foundation, 1969), a report of the Special Commission on the Social Sciences of the National Science Board, p. 7.

[26] See Erich Rothaker: *Einleitung in die Geisteswissenschaften* (Tübingen: Verlag won J.C.B. Mohr [Paul Siebeck], 1920; reprinted with detailed foreword (1930); E. Rothaker: *Logik und Systematik der Geisteswissenschaften* (Munich/Berlin: Druck und Verlag von R. Oldenbourg, 1926 – *Handbuch der Philiosophie*, ed. Alfred Baeumler and Manfred Schröter, numbers 6 and 7, collected in part 2, 1927; reprint, Bonn: H. Bouvier & Co. Verlag, 1947), esp. pp. 4–16.

Also Albrecht Timm: *Einführung in die Wissenschaftsgeschichte* (Munich: Wilhelm Fink Verlag, 1973), esp. pp. 37–48 and 137–140; Beat Sitter: *Die Geisteswissenschaften und ihre Bedeutung für unsere Zukunft* ([n.p.]: Schweizerische Volksbank, 1977), esp. pp. 13–17; Wolfgang Laskowski (ed.): *Geisteswissenschaft und Naturwissenschaft: Ihre Bedeutung für den Menschen von Heute* (Berlin: Verlag Walter de Gruyter & Co., 1970); Wolfram Krömer & Osmund Menghin (eds.): *Die Geisteswissenschaften stellen sich vor* (Innsbruck: Kommissionsverlag der Österreichischen Kommissionsbuchhandlung, 1983 – Veröffentlichungen der Universität Innsbruck, 137); Hans-Henrick Krummacher (ed.): *Geisteswissenschaften – wozu?: Beispiele ihrer Gegenstände uud ihrer Fragen* (Stuttgart: Franz Steiner Verlag, 1988); Erich Rothacker: *Einleitung in die Geisteswissenschaften* (Tübingen: Verlag von J.C.B. Mohr [Paul Siebeck], 1920; reprint, with detailed foreword, 1930); E. Rothaker: *Logik und Systematik der Geisteswissenschaften* (Munich/Berlin:

Druck uud Verlag won R. Oldenbourg, 1926 – *Handbuch der Philosophie*, ed. Alfred Baeumler & Manfred Schröter, nos. 6–7, 1927; reprint, Bonn: H. Bouvier & Co. Verlag, 1947), esp. pp. 4–16. See also L.H. Adolph Geck: "Über das Eindringen des Wortes 'sozial' in die Deutsche Sprache," *Soziale Welt*, 1962, **12**: 305–339.

[27] For a more recent historical study, including the usage of Geisteswissenschaften prior to the translation of Mill, see Alwin Diemer: "Die Differenzierung der Wissenschaften in die Natur- und die Geisteswissenschaften und die Begründung der Geisteswissenschaften als Wissenschaft," pp. 174–223 (esp. pp. 181–193) of A. Diemer (ed.): *Beiträge zur Entwicklung der Wissenschaftstheorie im 19. Jahrhundert* (Meisenheim am Glan: Verlag Anton Hain, 1968 – Studien zur Wissenschaftstheorie, vol. 1); A. Diemer: "Geisteswissenschaften," pp. 211–215 of Joachim Ritter (ed.): *Historisches Wörterbuch der Philiosophie*, vol. 3 (Basel/Stuttgart: Schwabe & Co. Verlag, 1974).

On Dilthey, see H.P. Richman: *Wilhelm Dilthey: Pioneer of the Human Studies* (Berkeley/Los Angeles/London: University of California Press, 1979), esp. pp. 58–73; and H.P. Rickman: *Dilthey Today: A Critical Appraisal of the Contemporary Relevance of his Work* (New York/Westport [Conn.]/London: Greenwood Press, 1988 – Contributions in Philosophy, no 35.), esp. pp. 79–82. In the latter (p. 80), Rickman errs in saying that Dilthey "introduced the term *Geisteswissenschaften* as a translation of J.S. Mill's 'moral sciences'"; as I have mentioned, J. Schiel did this in 1849 in his German version of Mill's *System of Logic*.

See also Wilhelm Dilthey: *Einleitung in die Geisteswissenschaften*, vol. 1 (Leipzig: Verlag von Dunker & Humblot, 1883), esp. pp. 5–7: this work is reprinted in Dilthey's *Gesammelte Schriften*, vol. 1 (Leipzig/Berlin: Verlag von B.G. Teubner, 1922; reprint, Stuttgart: B.G. Teubner Verlagsgesellschaft; Göttingen: Vandenhoeck & Ruprecht, 1959, 1962), see esp. pp. 4–6; there are a number of translations including Louis Sauzin (trans.): *Introduction à l'étude des sciences humaines* (Paris: Presses Universitaires de France, 1942), esp. pp. 13–15; Ramon J. Betanzos (trans.): *Introduction to the Human Sciences* (Detroit: Wayne State University Press, 1988), esp. pp. 77–79, also pp. 31–33; Michael Neville (trans.): *Introduction to the Human Sciences*, ed. Rudolf A. Makkreel & Frithjof Rodi (Princeton: Princeton Unviersity Press, 1989 – Selected Works, vol. 1), esp. pp. 56–58. See also Rothaker: *Logik und Systematik* (n. 26 supra), p. 6.

[28] John Stuart Mill: *Die inductive Logik*, trans. J. Schiel (Braunschweig: Verlag von Friedrich Vieweg & Sohn, 1849). This volume is quite rare; I have not been able to consult it directly. A second edition bears an enlarged title, J.S. Mill: *System der deductiven und inductiven Logik*, 2 vols. (Braunschweig: Druck und Verlag von Friedrich Vieweg und Sohn, 1862–1863); see esp. vol. 2, pp. 433, 437–438.

[29] John Stuart Mill: *System der deductiven und inductiven Logik*, trans. Theodor Gomperz, vol. 3 (Leipzig: Fues's Verlag [R. Reisland], 1873 – Gesammelte Werke, vol. 4), esp. pp. 229, 233–234.

[30] A. Diemer: "Die Differenzierung," (n. 27 supra), pp. 183–187, and "Geisteswissenschaften," p. 211.

[31] Hermann von Helmholz: "Über das Verhältnis der Naturwissenschaften zur Gesamtheit der Wissenschaften," *Philosophische Vorträge uud Aufsätze*, ed. Herbert Hörz & Siegfried Wollgast (Berlin: Akademie-Verlag, 1971), pp. 79–108; Hermann von Helmholz: *Das Denken in der Naturwissenschaft* (Darmstadt: Wissenschaftliche Buchgesellschaft, 1968), pp. 1–29; trans. Russell Kahl & H.W. Eve, "The Relation of the Natural Sciences to

Science in General," *Selected Writings of Hermann von Helmholz*, ed. Russell Kahl (Middletown [Conn.]: Wesleyan University Press, 1971), pp. 122–143. On this topic, see David E. Leary: "Telling Likely Stories: The Rhetoric of the New Psychology, 1880–1920," *Journal of the History of the Behavioral Sciences*, 1987, **23**: 315–331.

[32] H.A. Hodges: *The Philosophy of Wilhelm Dilthey* (London: Routledge & Kegan Paul, 1952; reprint, Westport [Conn.]: Greenwood Press, 1974), esp. pp. xxi–xxiii; Michael Ermarth: *Wilhelm Dilthey: The Critique of Historical Reason* (Chicago/London: The University of Chicago Press, 1978), esp. pp. 94–108, 359–360; H.P. Rickman: *Dilthey Today* (n. 27 supra), esp. pp. 79–82; Erich Rothacker (n. 26 supra), esp. pp. 253–277.

[33] Cf. Rudolf A. Makkreel: *Dilthey: Philosopher of the Human Studies* (Princeton: Princeton University Press, 1975), esp. pp. 35–44; H.P. Rickman: *Wilhelm Dilthey: Pioneer* (n. 27 supra), esp. pp. 58–73; and the works cited in nn. 2 and 7 supra.

INDEX